APPLIED
PARALLEL COMPUTING

APPLIED
PARALLEL COMPUTING

Yuefan Deng

Stony Brook University, USA

 **World Scientific**

NEW JERSEY • LONDON • SINGAPORE • BEIJING • SHANGHAI • HONG KONG • TAIPEI • CHENNAI

Published by

World Scientific Publishing Co. Pte. Ltd.

5 Toh Tuck Link, Singapore 596224

USA office: 27 Warren Street, Suite 401-402, Hackensack, NJ 07601

UK office: 57 Shelton Street, Covent Garden, London WC2H 9HE

British Library Cataloguing-in-Publication Data

A catalogue record for this book is available from the British Library.

APPLIED PARALLEL COMPUTING

ISBN 978-981-4307-60-4

Typeset by Stallion Press

Email: enquiries@stallionpress.com

Printed in Singapore.

PREFACE

This manuscript, Applied Parallel Computing, gathers the core materials from a graduate course of similar title I have been teaching at Stony Brook University for 23 years, and from a summer course I taught at the Hong Kong University of Science and Technology in 1995, as well as from multiple month-long and week-long training sessions at the following institutions: HKU, CUHK, HK Polytechnic, HKBC, the Institute of Applied Physics and Computational Mathematics in Beijing, Columbia University, Brookhaven National Laboratory, Northrop-Grumman Corporation, and METU in Turkey, KISTI in Korea.

The majority of the attendees are advanced undergraduate and graduate science and engineering students requiring skills in large-scale computing. Students in computer science, economics, and applied mathematics are common to see in classes, too.

Many participants of the above events contributed to the improvement and completion of the manuscript. My former and current graduate students, J. Braunstein, Y. Chen, B. Fang, Y. Gao, T. Polishchuk, R. Powell, and P. Zhang have contributed new materials from their theses. Z. Lou, now a graduate student at the University of Chicago, edited most of the manuscript and I wish to include him as a co-author.

Supercomputing experiences super development and is still evolving. This manuscript will evolve as well.

CONTENTS

CHAPTER 1

INTRODUCTION

1.1. Definition of Parallel Computing

The US Department of Energy defines parallel computing as:

"simultaneous processing by more than one processing unit on a single application".[1]

It is the ultimate approach for a large number of large-scale scientific, engineering, and commercial computations.

Serial computing systems have been with us for more than five decades since John von Neumann introduced digital computing in the 1950s. A serial computer refers to a system with one central processing unit (CPU) and one memory unit, which may be so arranged as to achieve efficient referencing of data in the memory. The overall speed of a serial computer is determined by the execution clock rate of instructions and the bandwidth between the memory and the instruction unit.

To speed up the execution, one would need to either increase the clock rate or reduce the computer size to lessen the signal travel time.

Both are approaching the fundamental limit of physics at an alarming pace. Figure 1.1 also shows that as we pack more and more transistors on a chip, more creative and prohibitively expensive cooling techniques are required as shown in Fig. 1.2. At this time, two options survive to allow sizable expansion of the bandwidth. First, memory interleaving divides memory into banks that are accessed independently by the multiple channels. Second, memory caching divides memory into banks of hierarchical accessibility by the instruction unit, e.g. a small amount of fast memory and large amount of slow memory. Both of these efforts increase CPU speed, but only marginally, with frequency walls and memory bandwidth walls.

Vector processing is another intermediate step for increasing speed. One central processing unit controls several (typically, around a dozen) vector

---

[1]http://www.nitrd.gov/pubs/bluebooks/1995/section.5.html.

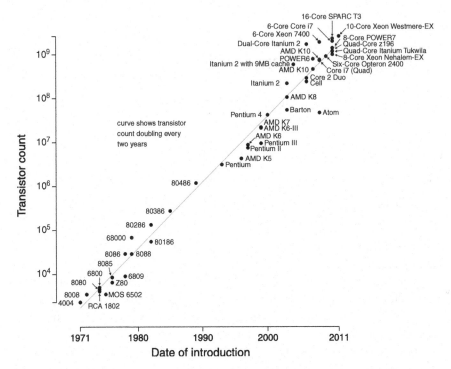

Fig. 1.1.   Microprocessor transistor counts 1971–2011 and Moore's Law.
Source: Wgsimon on Wikipedia.

units that can work simultaneously. The advantage is the simultaneous execution of instructions in these vector units for several factors of speed up over serial processing. However, there exist some difficulties. Structuring the application codes to fully utilize vector units and increasing the scale of the system limit improvement.

Obviously, the difficulty in utilizing a computer — serial or vector or parallel — increases with the complexity of the systems. Thus, we can safely make a list of the computer systems according to the level of difficulty of programming them:

(1) Serial computers.
(2) Vector computers with well-developed software systems, especially compilers.
(3) Shared-memory computers whose communications are handled by referencing to the global variables.
(4) Distributed-memory single-instruction multiple-data computers with the data parallel programming models; and

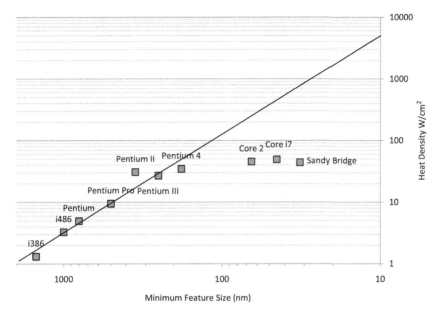

Fig. 1.2. Microprocessors' dissipated heat density vs. feature size.

(5) Distributed-memory multiple-instruction multiple-data systems whose communications are carried out explicitly by message passing.

On the other hand, if we consider their raw performance and flexibility, then the above computer systems will be rearranged thus: distributed-memory multiple-instruction multiple-data system, distributed-memory single-instruction multiple-data system, shared-memory, vector, and finally the serial computers. Indeed, "No free lunch theorem"[2] applies rigorously here.

Programming system of a distributed-memory multiple-instruction multiple-data system will certainly burn more human neurons but one can get many bigger problems solved more rapidly.

Parallel computing has been enabling many scientific and engineering projects as well as commercial enterprises such as internet services and the recently developed cloud computing technology. It is easy to predict that parallel computing will continue to play essential roles in aspects of human activities including entertainment and education, but it is difficult to imagine where parallel computing will lead us to.

---

[2]D.H. Wolpert and W.G. Macready, "No free lunch theorems for optimization", *IEEE Transactions on Evolutionary Computation* **1** (1997) 67.

Thomas J. Watson, Chairman of the Board of International Business Machines, was quoted, or misquoted, for a 1943 statement: "*I think there is a world market for maybe five computers.*" Watson may be truly wrong as the world market is not of five computers and, rather, the whole world needs only one big parallel computer.

World's Top500 supercomputers are ranked bi-annually in terms of their LINPACK performance. Table 1.3 shows the Top10 such computers in June 2011.

1.2. Evolution of Computers

It is the increasing speed of parallel computers that defines their tremendous value to the scientific community. Table 1.1 defines computer speeds while Table 1.2 illustrates the times for solving representative medium sized and grand challenge problems.

Table 1.1.   Definitions of computing speeds.

Speeds	Floating-point operations per second	Representative computer
1 Flop	$10^0 = 1$	A fast human
1Kflops	$10^3 = 1$ Thousand	
1Mflops	$10^6 = 1$ Million	
1Gflops	$10^9 = 1$ Billion	VPX 220 (Rank #250 in 1993); A laptop in 2010
1Tflops	$10^{12} = 1$ Trillion	ASCI Red (Rank #1 in 1997)
1Pflops	$10^{15} = 1$ Quadrillion	IBM Roadrunner (Rank #1 in 2008); Cray XT5 (Rank #2 in 2008); 1/8 Fujitsu K Computer (Rank #1 in 2011)
1Eflops	$10^{18} = 1$ Quintillion	Expected in 2018
1Zflops	$10^{21} = 1$ Sextillion	
1Yflops	$10^{24} =$ Septillion	

Table 1.2.   Time scales for solving medium sized and grand challenge problems.

Computer	Moderate problems	Grand challenge problems	Applications
Sub-Petaflop	N/A	$O(1)$ Hours	Protein folding, QCD, and Turbulence
Tereflop	$O(1)$ Seconds	$O(10)$ Hours	Weather
1,000 Nodes Beowulf Cluster	$O(1)$ Minutes	$O(1)$ Weeks	2D CFD, Simple designs
High-end Workstation	$O(1)$ Hours	$O(10)$ Years	
PC with 2 GHz Pentium	$O(1)$ Days	$O(100)$ Years	

Table 1.3. World's Top10 supercomputers in June 2011.

	Vendor	Year	Computer	Rmax (Tflops)	Cores	Site	Country
1	Fujitsu	2011	K computer	8,162	548,352	RIKEN Advanced Institute for Computational Science	Japan
2	NUDT	2010	Tianhe-1A	2,566	186,368	National Supercomputing Center in Tianjin	China
3	Cray	2009	Jaguar Cray XT5-HE	1,759	224,162	DOE/SC/Oak Ridge National Laboratory	USA
4	Dawning	2010	Nebulae Dawning Cluster	1,271	120,640	National Supercomputing Centre in Shenzhen	China
5	NEC/HP	2010	TSUBAME 2.0 HP Cluster Platform 3,000SL	1,192	73,278	GSIC Center, Tokyo Institute of Technology	Japan
6	Cray	2011	Cielo Cray XE6	1,110	142,272	DOE/NNSA/ LANL/SNL	USA
7	SGI	2011	Pleiades SGI Altix	1,088	111,104	NASA/Ames Research Center/NAS	USA
8	Cray	2010	Hopper Cray XE6	1,054	153,408	DOE/SC/LBNL/ NERSC	USA
9	Bull SA	2010	Tera-100 Bull Bullx	1,050	138,368	Commissariat a l'Energie Atomique	France
10	IBM	2009	Roadrunner IBM BladeCenter	1,042	122,400	DOE/NNSA/ LANL	USA

A grand challenge is a fundamental problem in science or engineering, with a broad application, whose solution would be enabled by the application of the high performance computing resources that could become available in the near future. For example, a grand challenge problem in 2011, which would require 1,500 years to solve on a high-end workstation, could be solved on the latest faster K Computer in a few hours.

Measuring computer speed is itself an evolutionary process. Instructions per-second (IPS) is a measure of processor speed. Many reported speeds have represented peak rates on artificial instructions with few branches or memory referencing varieties, whereas realistic workloads typically lead to significantly lower speeds. Because of these problems of inflating speeds, researchers created standardized tests such as SPECint as an attempt to measure the real effective performance in commonly used applications.

In scientific computations, the LINPACK Benchmarks are used to measure a system's floating point computing speed. It measures how fast a computer solves a dense N-dimensional system of linear equations commonly appearing in engineering. The solution is obtained by Gaussian elimination with partial pivoting. The result is reported in millions of floating point operations per second.

Figure 1.3 demonstrates Moore's law in action for computer speeds over the past 50 years while Fig. 1.4 shows the evolution of computer architectures. During the 20 years, since 1950s, mainframes were the main computers and several users shared one processor. During the next 20 years, since 1980s, workstations and personal computers formed the majority of the computers where each user had a *personal* processor. During the next unknown number of years (certainly more than 20), since 1990s, parallel computers have been, and will most likely continue to be, dominating the

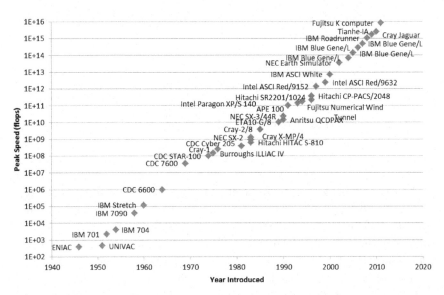

Fig. 1.3. The peak performance of supercomputers.
Source: Top500.org and various other sources, prior to 1993.

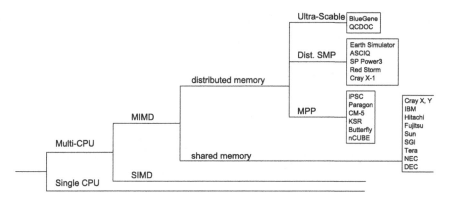

Fig. 1.4.   Evolution of computer architectures.

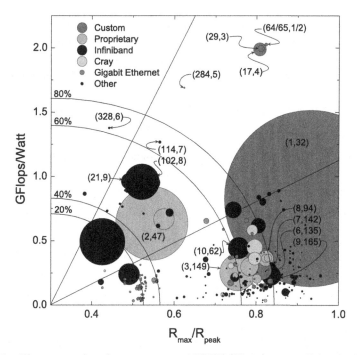

Fig. 1.5.   The scatter plot of supercomputers' LINPACK and power efficiencies in 2011.

user space where a single user will control many processors. Figure 1.5 shows the 2011's supercomputer's LINPACK and power efficiencies.

It is apparent from these figures that parallel computing is of enormous value to scientists and engineers who need computing power.

1.3. An Enabling Technology

Parallel computing is a fundamental and irreplaceable technique used in today's science and technology, as well as manufacturing and service industries. Its applications cover a wide range of disciplines:

- Basic science research, including biochemistry for decoding human genetic information as well as theoretical physics for understanding the interactions of quarks and possible unification of all four forces.
- Mechanical, electrical, and materials engineering for producing better materials such as solar cells, LCD displays, LED lighting, etc.
- Service industry, including telecommunications and the financial industry.
- Manufacturing, such as design and operation of aircrafts and bullet trains.

Its broad applications in oil exploration, weather forecasting, communication, transportation, and aerospace make it a unique technique for national economical defense. It is precisely this uniqueness and its lasting impact that defines its role in today's rapidly growing technological society.

Parallel computing research's main concerns are:

(1) Analysis and design of

    a. Parallel hardware and software systems.

    b. Parallel algorithms.

    c. Parallel programs.

(2) Development of applications in science, engineering, and commerce.

The processing power offered by parallel computing, growing at a rate of orders of magnitude higher than the impressive 60% annual rate for microprocessors, has enabled many projects and offered tremendous potential for basic research and technological advancement. Every computational science problem can benefit from parallel computing.

Supercomputing power has been energizing the science and technology community and facilitating our daily life. As it has been, supercomputing development will stimulate the birth of new technologies in hardware, software, algorithms while enabling many other areas of scientific discoveries including maturing the 3rd and likely the 4th paradigms of scientific research. This new round of Exascale computing will bring unique excitement in lifting energy and human efficiencies in adopting and adapting electronic technologies.

1.4. Cost Effectiveness

The total cost of ownership of parallel computing technologies includes:

(1) Purchasing cost.
(2) Operating cost including maintenance and utilities.
(3) Programming cost in terms of added training for users.

In fact, there is an additional cost of time, i.e., cost of delay in the technology deployment due to lack of adequate knowledge in realizing its potential. How to make a sound investment in time and money on adopting parallel computing is the complex issue.

Apart from business considerations that parallel computing is a cost-effective technology, it can be the only option for the following reasons:

(1) To improve the absolute response time.
(2) To study problems of absolutely largest spatial size at the highest spatial resolution.
(3) To study problems of absolutely largest time scale at the highest temporal resolutions.

The analytical methods used to solve scientific and engineering problems were driven "out of fashion" several decades ago due to the growing complexity of these problems. Solving problems numerically — on serial computers available at the time — had been quite attractive for 20 years or so, starting in the 1960s. This alternative of solving problems with serial computers was quite successful, but became obsolete with the gradual advancement of a new parallel computing technique available only since the early 1990s. Indeed, parallel computing is the wave of the future.

1.4.1. *Purchasing costs*

Hardware costs are the major expenses for running any supercomputer center. Interestingly, the hardware costs per unit performance have been decreasing steadily while those of operating a supercomputer center including utilities to power them up and cool them off, and administrator's salaries have been increasing steadily. 2010 marked the turning point when the hardware costs were less than the operating costs.

Table 1.4 shows a list of examples of computers that demonstrates how drastically performance has increased and price has decreased. The "cost per Gflops" is the cost for a set of hardware that would theoretically operate

Table 1.4.   Hardware cost per Gflops at different times.

Date	Cost per Gflops	Representative technology
1961	1.1×10^{12}	IBM 1620 (costing $64K)
1984	1.5×10^7	Cray Y-MP
1997	3.0×10^4	A Beowulf Cluster
2000	3.0×10^3	Workstation
2005	1.0×10^3	PC Laptop
2007	4.8×10^1	Microwulf Cluster
2011	1.8×10^0	HPU4Science Cluster
2011	1.2×10^0	K Computer power cost

at one billion floating-point operations per second. During the era when no single computing platform was able to achieve one Gflops, this table lists the total cost for multiple instances of a fast computing platform whose speed sums to one Gflops. Otherwise, the least expensive computing platform able to achieve one Gflops is listed.

1.4.2. *Operating costs*

Most of the operating costs involve powering up the hardware and cooling it off. The latest (June 2011) Green500[3] list shows that the most efficient Top500 supercomputer runs at 2097.19 Mflops per watt, i.e. an energy requirement of 0.5 W per Gflops. Operating such a system for one year will cost 4 KWh per Gflops. Therefore, the lowest annual power consumption of operating the most power-efficient system of 1 Pflops is 4,000,000 KWh. The energy cost on Long Island, New York, in 2011 is $0.25 per KWh, so the annual energy monetary cost of operating a 1 Pflops system is $1M. Quoting the same Green500 list, the least efficient Top500 supercomputer runs at 21.36 Mflops per watt, or nearly 100 times less power efficient than the most power efficient system mentioned above. Thus, if we were to run such a system of 1 Pflops, the power cost is $100M per year. In summary, the annual costs of operating the most and the least power-efficient systems among the Top500 supercomputer of 1 Pflops in 2011 are $1M and $100M respectively, and the median cost of $12M as shown by Table 1.5.

---

[3]http://www.green500.org

Table 1.5. Supercomputer operating cost estimates.

Computer	Green500 rank	Mflops/W	1 Pflops operating cost
IBM BlueGene/Q	1	2097.19	$1M
Bullx B500 Cluster	250	169.15	$12M
PowerEdge 1850	500	21.36	$100M

1.4.3. *Programming costs*

The level of programmers investment in utilizing a computer, serial or vector or parallel, increases with the complexity of the system, naturally.

Given the progress in parallel computing research, including software and algorithm development, we expect the difficulty in employing most supercomputer systems to reduce gradually. This is largely due to the popularity and tremendous values of this type of complex systems. This manuscript also attempts to make programming parallel computers enjoyable and productive.

CHAPTER 2

PERFORMANCE METRICS AND MODELS

Measuring the performance of a parallel algorithm is somewhat tricky due to the inherent complication of the relatively immature hardware system or software tools or both, and the complexity of the algorithms, plus the lack of proper definition of timing for different stages of a certain algorithm. However, we will examine the definitions for speedup, parallel efficiency, overhead, and scalability, as well as Amdahl's "law" to conclude this chapter.

2.1. Parallel Activity Trace

It is cumbersome to describe, let alone analyze, parallel algorithms. We have introduced a new graphic system, which we call parallel activity trace (PAT) graph, to illustrate parallel algorithms. The following is an example created with a list of conventions we establish:

(1) A 2D Cartesian coordinate system is adopted to depict the graph with the horizontal axis for wall clock time and the vertical axis for the "ranks" or IDs of processors or nodes or cores depending on the level of details that we wish to examine the activities.

(2) A horizontal green bar is used to indicate a local serial computation on a specific processor or computing unit. Naturally, the two ends of the bar indicate the starting and ending times of the computation and thus the length of the bar shows the amount of time the underlying serial computation takes. Above the bar, one may write down the function being executed.

(3) A red wavy line indicates the processor is sending a message. The two ends of the line indicate the starting and ending times of the message sending process and thus the length of the line shows the amount of time for sending a message to one or more processors. Above the line, one may write down the "ranks" or IDs of the receiver of the message.

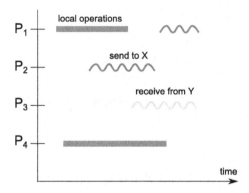

Fig. 2.1.   A PAT graph for illustrating the symbol conventions.

(4) A yellow wavy line indicates the processor is receiving a message. The two ends of the line indicate the starting and ending times of the message receiving process and thus the length of the line shows the amount of time for receiving a message. Above the line, one may write down the "ranks" or IDs of the sender of the message.

(5) An empty interval signifies the fact that the processing unit is idle.

We expect the PAT graph (Fig. 2.1) to record vividly, *a priori* or posterior, the time series of multi-processor activities in the parallel computer. The fact that many modern parallel computers can perform message passing and local computation simultaneously may lead to overlapping of the above lines. With such a PAT graph, one can easily examine the amount of local computation, amount of communication, and load distribution, etc. As a result, one can visually consider, or obtain guide for, minimization of communication and load imbalance. We will follow this simple PAT graphic system to describe our algorithms and hope this little "invention" of ours would aid in parallel algorithm designs.

2.2. Speedup

Let $T(1,N)$ be the time required for the best serial algorithm to solve problem of size N on one processor and $T(P,N)$ be the time for a given parallel algorithm to solve the same problem of the same size N on P processors. Thus, speedup is defined as:

$$S(P,N) = \frac{T(1,N)}{T(P,N)}. \tag{2.1}$$

Normally, $S(P, N) \leq P$. Ideally, $S(P, N) = P$. Rarely, $S(P, N) > P$; this is known as *super speedup*.

For some memory intensive applications, super speedup may occur for some small N because of memory utilization. Increase in N also increases the amount of memory, which will reduce the frequency of the swapping, hence largely increasing the speedup. The effect of memory increase will fade away when N becomes large. For these kinds of applications, it is better to measure the speedup based on some P_0 processors rather than one. Thus, the speedup can be defined as:

$$S(P, N) = \frac{P_0 T(P_0, N)}{T(P, N)}. \tag{2.2}$$

Most of the time, it is easy to speedup large problems than small ones.

2.3. Parallel Efficiency

Parallel efficiency is defined as:

$$E(P, N) = \frac{T(1, N)}{T(P, N)P} = \frac{S(P, N)}{P}. \tag{2.3}$$

Normally, $E(P, N) \leq 1$. Ideally, $E(P, N) = 1$. Rarely is $E(P, N) > 1$. It is generally acceptable to have $E(P, N) \sim 0.6$. Of course, it is problem dependent.

A linear speedup occurs when $E(P, N) = c$, where c is independent of N and P.

Algorithms with $E(P, N) = c$ are called scalable.

2.4. Load Imbalance

If processor i spends T_i time doing useful work (Fig. 2.2), the total time spent working by all processors is $\sum_{i=1}^{P-1} T_i$ and the average time a processor spends working is:

$$T_{\text{avg}} = \frac{\sum_{i=0}^{P-1} T_i}{P}. \tag{2.4}$$

The term $T_{\text{max}} = \max\{T_i\}$ is the maximum time spent by any processor, so the total processor time is PT_{max}. Thus, the parameter called the load imbalance ratio is given by:

$$I(P, N) = \frac{PT_{\text{max}} - \sum_{i=0}^{P-1} T_i}{\sum_{i=0}^{P-1} T_i} = \frac{T_{\text{max}}}{T_{\text{avg}}} - 1. \tag{2.5}$$

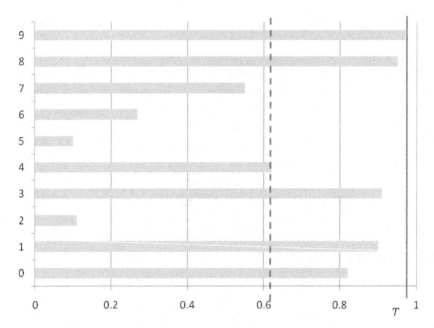

Fig. 2.2.   Computing time distribution: Time t_i on processor i.

Remarks

➤ $I(P, N)$ is the average time wasted by each processor due to load imbalance.

➤ If $T_i = T_{\max}$ for every i, then $I(P, N) = 0$, resulting in a complete load balancing.

➤ The slowest processor $T_{\max}$ can mess up the entire team. This observation shows that slave-master scheme is usually very inefficient because of the load imbalance issue due to slow master processor.

2.5. Granularity

The size of the sub-domains allocated to the processors is called the granularity of the decomposition. Here is a list of remarks:

➤ Granularity is usually determined by the problem size N and computer size P.

➤ Decreasing granularity usually increases communication and decreases load imbalance.

➤ Increasing granularity usually decreases communication and increases load imbalance.

2.6. Overhead

In all parallel computing, it is the communication and load imbalance overhead that affects the parallel efficiency. Communication costs are usually buried in the processor active time. When a co-processor is added for communication, the situation becomes trickier.

We introduce a quantity called load balance ratio. For an algorithm using P processors, at the end of one logical point (e.g. a synchronization point), processor i is busy, either computing or communicating, for t_i amount of time.

Let $t_{\max} = \max\{t_i\}$ and the time that the entire system of P processors spent for computation or communication is $\sum_{i=1}^{P} t_i$. Finally, let the total time that all processors are occupied (by computation, communication, or being idle) be $Pt_{\max}$. The ratio of these two is defined as the load balance ratio:

$$L(P, N) = \frac{\sum_{i=1}^{P} T_i}{PT_{\max}} = \frac{T_{\text{avg}}}{T_{\max}}. \tag{2.6}$$

Obviously, if $T_{\max}$ is close to T_{avg}, the load must be well balanced, so $L(P, N)$ approaches one. On the other hand, if $T_{\max}$ is much larger than T_{avg}, the load must not be balanced, so $L(P, N)$ tends to be small.

Two extreme cases are

➤ $L(P, N) = 0$ for total imbalance.
➤ $L(P, N) = 1$ for perfect balance.

$L(P, N)$ only measures the percentage of the utilization during the system "up time," which does not care what the system is doing. For example, if we only keep one of the $P = 2$ processors in a system busy, we get $L(2N) = 50\%$, meaning we achieved 50% utilization. If $P = 100$ and one is used, then $L(100, N) = 1\%$, which is badly imbalanced.

Also, we define the load imbalance ratio as $1 - L(P, N)$. The overhead is defined as

$$H(P, N) = \frac{P}{S(P, N)} - 1. \tag{2.7}$$

Normally, $S(P, N) \leq P$. Ideally, $S(P, N) = P$. A linear speedup means that $S(P, N) = cP$ where c is independent of N and P.

2.7. Scalability

First, we define two terms: Scalable algorithm and quasi-scalable algorithm. A scalable algorithm is defined as those whose parallel efficiency $E(P, N)$ remains bounded from below, i.e. $E(P, N) \geq E_0 > 0$, when the number of processors $P \to \infty$ at fixed problem size.

More specifically, those which can keep the efficiency constant when the problem size N is kept constant are called strong scalable, and those which can keep the efficiency constant only when N increases along with P are called weak scalable.

A quasi-scalable algorithm is defined as those whose parallel efficiency $E(P, N)$ remains bounded from below, i.e. $E(P, N) \geq E_0 > 0$, when the number of processors $P_{\min} < P < P_{\max}$ at fixed problem size. The interval $P_{\min} < P < P_{\max}$ is called scaling zone.

Very often, at fixed problem size $N = N(P)$, the parallel efficiency decreases monotonically as the number of processors increase. This means that for sufficiently large number of processors the parallel efficiency tends to vanish. On the other hand, if we fix the number of processors, the parallel efficiency usually decreases as the problem size decreases. Thus, very few algorithms (aside from the embarrassingly parallel algorithm) are scalable, while many are quasi-scalable. Two major tasks in designing parallel algorithms are to maximize E_0 and the scaling zone.

2.8. Amdahl's Law

Suppose a fraction f of an algorithm for a problem of size N on P processors is inherently serial and the remainder is perfectly parallel, then assume:

$$T(1, N) = \tau. \tag{2.8}$$

Thus,

$$T(P, N) = f\tau + (1 - f)\tau/P. \tag{2.9}$$

Therefore,

$$S(P, N) = \frac{1}{f + (1 - f)/P}. \tag{2.10}$$

This indicates that when $P \to \infty$, the speedup $S(P, N)$ is bounded by $1/f$. It means that the maximum possible speedup is finite even if $P \to \infty$.

CHAPTER 3

HARDWARE SYSTEMS

A serial computer with one CPU and one chunk of memory while ignoring the details of its possible memory hierarchy and some peripherals, needs only two parameters/properties to describe itself: Its CPU speed and its memory size.

On the other hand, five or more properties are required to characterize a parallel computer:

(1) Number and properties of computer processors;
(2) Network topologies and communication bandwidth;
(3) Instruction and data streams;
(4) Processor-memory connectivity;
(5) Memory size and I/O.

3.1. Node Architectures

One of the greatest advantages of a parallel computer is the simplicity in building the processing units. Conventional, off-the-shelf, and mass-product processors are normally used in contrast to developing special-purpose processors such as those for Cray processors and for IBM mainframe CPUs.

In recent years, the vast majority of the designs are centered on four of the processor families: Power, AMD x86-64, Intel EM64T, and Intel Itanium IA-64. These four together with Cray and NEC families of vector processors are the only architectures that are still being actively utilized in the high-end supercomputer systems. As shown in the Table 3.1 (constructed with data from top500.org for June 2011 release of Top500 supercomputers) 90% of the supercomputers use x86 processors.

Ability of the third-party organizations to use available processor technologies in original HPC designs is influenced by the fact that the companies that produce end-user systems themselves own PowerPC, Cray

Table 3.1.   Top500 supercomputers processor shares in June 2011.

Processor family	Count	Share %	Rmax sum (GF)
Power	45	9.00%	6,274,131
NEC	1	0.20%	122,400
Sparc	2	0.40%	8,272,600
Intel IA-64	5	1.00%	269,498
Intel EM64T	380	76.00%	31,597,252
AMD x86_64	66	13.20%	12,351,314
Intel Core	1	0.20%	42,830
Totals	**500**	**100%**	**58,930,025**

and NEC processor families. AMD and Intel do not compete in the end-user HPC system market. Thus, it should be expected that IBM, Cray and NEC would continue to control the system designs based on their own processor architectures, while the efforts of the other competitors will be based on processors provided by Intel and AMD.

Currently both companies, Intel and AMD, are revamping their product lines and are transitioning their server offerings to the quad-core processor designs. AMD introduced its Barcelona core on 65 nm manufacturing process as a competitor to the Intel Core architecture that should be able to offer a comparable to Core 2 instruction per clock performance, however the launch has been plagued by delays caused by the difficulties in manufacturing sufficient number of higher clocked units and emerging operational issues requiring additional bug fixing that so far have resulted in sub-par performance. At the same time, Intel enhanced their product line with the Penryn core refresh on a 45 nm process featuring ramped up the clock speed, optimized execution subunits, additional SSE4 instructions, while keeping the power consumption down within previously defined TDP limits of 50 W for the energy-efficient, 80 W for the standard and 130 W for the high-end parts. According to the roadmaps published by both companies, the parts available in 2008 consisting up to four cores on the same processor, die with peak performance per core in the range of 8–15 Gflops on a power budget of 15–30 W and 16–32 Gflops on a power budget of 50–68 W. Due to the superior manufacturing capabilities, Intel is expected to maintain its performance per watt advantage with top Penryn parts clocked at 3.5 GHz or above, while the AMD Barcelona parts in the same power envelope was not expected to exceed 2.5 GHz clock speed until the second half of 2008 at best. The features of the three processor-families that power the top supercomputers in 2–11 are

Table 3.2. Power consumption and performance of processors in 2011.

Parameter	Fujitsu SPARC64 VIIIfx[1]	IBM power BQC	AMD opteron 6100 series
Core Count	8	16	8
Highest Clock Speed	2 GHz	1.6 GHz	2.3 Ghz (est)
L2 Cache	5 MB		4 MB
L3 Cache	N/A		12 MB
Memory Bandwidth	64 GB/s		42.7 GB/s
Manufacturing Technology	45 nm		45 nm
Thermal Design Power Bins	58 W		115 W
Peak Floating Point Rate	128 Gflops	205 Gflops	150 Gflops

given in Table 3.2 with the data collected from respective companies' websites.

3.2. Network Interconnections

As the processor's clock speed hits the limit of physics laws, the ambition of building a record-breaking computer relies more and more on embedding ever-growing numbers of processors into a single system. Thus, the network performance has to speedup along with the number of processors so as not to be the bottleneck of the system.

The number of ways of connecting a group of processors is very large. Experience indicates that only a few are optimal. Of course, each network exists for a special purpose. With the diversity of the applications, it does not make sense to say which one is the best in general. The structure of a network is usually measured by the following parameters:

➤ **Connectivity:** Multiplicity between any two processors.

➤ **Average distance:** The average of distances from a reference node to all other nodes in the topology. Such average should be independent of the choice of reference nodes in a well-designed network topology.

➤ **Diameter:** Maximum distance between two processors (in other words, the number of "hops" between two most distant processors).

➤ **Bisection bandwidth:** Number of bits that can be transmitted in parallel multiplied by the bisection width. The **bisection bandwidth** is defined as the minimum number of communication links that have to

---

[1]http://www.fujitsu.com/downloads/TC/090825HotChips21.pdf.

be removed to divide the network into two partitions **with an equal number** of processors.

3.2.1. *Topology*

Here are some of the topologies that are currently in common use:

(1) Multidimensional mesh or torus (Fig. 3.1–3.2).
(2) Multidimensional hypercube.
(3) Fat tree (Fig. 3.3).
(4) Bus and switches.
(5) Crossbar.
(6) Ω Network.

Practically, we call a 1D mesh an array and call a 1D torus a ring. Figures 3.1–3.4 illustrate some of the topologies. Note the subtle similarities and differences of the mesh and torus structure. Basically, a torus network can be constructed from a similar mesh network by wrapping the end points of every dimension of the mesh network. In practice, as in the IBM BlueGene systems, the physically constructed torus networks can be manipulated to

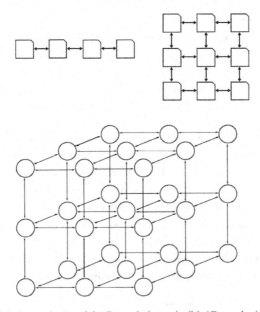

Fig. 3.1.   Mesh topologies: (a) 1D mesh (array); (b) 2D mesh; (c) 3D mesh.

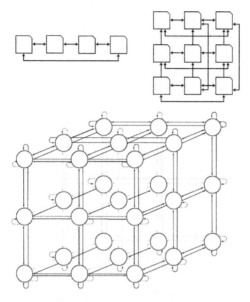

Fig. 3.2.   Torus topologies: (a) 1D torus (ring); (b) 2D torus; (c) 3D torus.

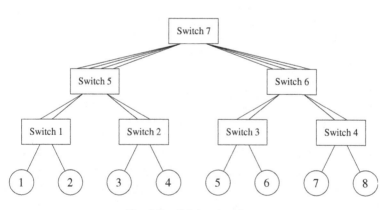

Fig. 3.3.   Fat tree topology.

retain the torus structure or to become a mesh network, at will, by software means.

Among those topologies, mesh, torus and fat tree are more frequently adopted in latest systems. Their properties are summarized in Table 3.3.

Mesh interconnects permit a higher degree of hardware scalability than one can afford by the fat tree network topology due to the absence of

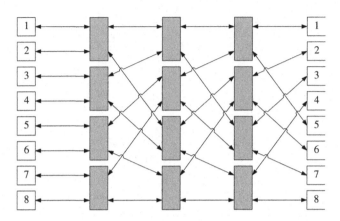

Fig. 3.4.   An $8 \times 8\,\Omega$ Network topology.

Table 3.3.   Properties of mesh, torus, and fat tree networks.

	Mesh or torus	Fat tree
Advantages	Fast local communication	Off-the-shelf
	Fault tolerant	Good remote communication
	Easy to scale	Economical
	Low latency	
Disadvantages	Poor remote communication	Fault sensitive
		Hard to scale up
		High latency

the external federated switches formed by a large number of individual switches. Thus, a large-scale cluster design with several thousand nodes has to necessarily contain twice that number of the external cables connecting the nodes to tens of switches located tens of meters away from nodes. At these port counts and multi-gigabit link speeds, it becomes cumbersome to maintain the quality of individual connections, which leads to maintenance problems that tend to accumulate over time.

As observed on systems with Myrinet and Infiniband interconnects, intermittent errors on a single cable may have a serious impact on the performance of the massive parallel applications by slowing down the communication due to the time required for connection recovery and data retransmission. Mesh network designs overcome this issue by implementing a large portion of data links as traces on the system backplanes and by aggregating several cable connections into a single bundle attached to a single socket, thus reducing the number of possible mechanical faults by an order of magnitude. These considerations are confirmed by a review of the changes for the Top10 supercomputers. In June 2011, there were four systems based on mesh networks and six based on fat tree networks, a much better ratio than the one for the all Top500 supercomputers, which leads to the conclusion that advanced interconnection networks such as torus are of a greater importance in scalable high end systems (Table 3.4).

Interestingly, SGI Altix ICE interconnect architecture is based on Infiniband hardware, typically used as a fat tree network in cluster applications, which is organized as a fat tree on a single chassis level and as a mesh for chassis to chassis connections. This confirms the observations about problems associated with using a fat tree network for large-scale system designs.

Advanced cellular network topology

Since the communication rates of a single physical channel are also limited by the clock speed of network chips, the practical way to boost bandwidth is to add the number of network channels in a computing node, that is, when applied to the mesh or torus network we discussed, this increases the number of dimensions in the network topology. However, the dimension cannot grow without restriction. That makes a cleverer design of network more desirable in top-ranked computers. Here, we present two advanced cellular network topology that might be the trend for the newer computers.

Table 3.4. Overview of the Top10 supercomputers in June 2011.

Rank	System	Speed (Tflops)	Processor family	Co-processor family	Interconnect	Interconnect topology
1	K Computer	8162	SPARC64	N/A	Tofu	6D Torus
2	Tianhe-1A	2566	Intel EM64T	nVidia GPU	Proprietary	Fat Tree
3	Cray XT5	1759	AMD x86_64	N/A	SeaStar	3D Torus
4	Dawning	1271	Intel EM64T	nVidia Tesla	Infiniband	Fat Tree
5	HP ProLiant	1192	Intel EM64T	nVidia GPU	Infiniband	Fat Tree
6	Cray XE6	1110	AMD x86_64	N/A	Gemini	3D Torus
7	SGI Altix ICE	1088	Intel EM64T	N/A	Infiniband	Fat Tree
8	Cray XE6	1054	AMD x86_64	N/A	Gemini	3D Torus
9	Bull Bullx	1050	AMD x86_64	N/A	Infiniband	Fat Tree
10	IBM BladeCenter	1042	AMD x86_64	IBM Power X Cell	Infiniband	Fat Tree

MPU networks

The Micro Processor Unit (MPU) network is a combination of two k-dimensional rectangular meshes of equal size, which are offset by $1/2$ of a hop along each dimension to surround each vertex from one mesh with a cube of 2^k neighbors from the other mesh. Connecting vertices in one mesh diagonally to their immediate neighbors in the other mesh and removing original rectangle mesh connections produces the MPU network. Figure 3.5 illustrates the generation of 2D MPU network by combining two 2D meshes. Constructing MPU network of three or higher dimension is similar. To complete wrap-around connections for boundary nodes, we apply the cyclic property of a symmetric topology in such a way that a boundary node encapsulates in its virtual multi-dimensional cube and connects to all vertices of that cube.

In order to see the advantages of MPU topology, we compare MPU topologies with torus in terms of key performance metrics as network diameter, bisection width/bandwidth and average distance. Table 3.5 lists the comparison of those under same dimension.

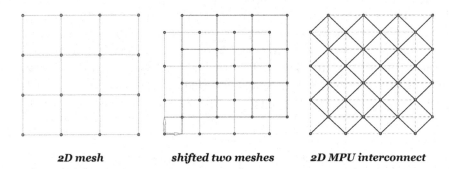

2D mesh   *shifted two meshes*   *2D MPU interconnect*

Fig. 3.5.   2D MPU generated by combining two shifted 2D meshes.

Table 3.5.   Analytical comparisons between MPU and torus networks.

Network of characters	MPU (n^k)	Torus (n^k)	Ratio of MPU to torus
Dimensionality	k	k	1
Number of nodes	$2n^k$	n^k	2
Node degree	2^k	2^k	$2^{k-1}k^{-1}$
Network diameter	n	$nk/2$	$2k^{-1}$
Bisection width	$2^k n^{k-1}$	$2n^{k-1}$	2^{k-1}
Bisection bandwidth	pn^{k-1}	$pn^{k-1}k^{-1}$	k
Number of wires	$2^k n^k$	kn^k	$2^k k^{-1}$

MSRT networks

A 3D Modified Shift Recursive Torus (MSRT) network is a 3D hierarchical network consisting of massive nodes that are connected by a basis 3D torus network and a 2D expansion network. It is constructed by adding bypass links to a torus. Each node in 3D MSRT has eight links to other nodes, of which six are connected to nearest neighbors in a 3D torus and two are connected to bypass neighbors in the 2D expansion network. The MSRT achieves shorter network diameter and higher bisection without increasing the node degree or wiring complexity.

To understand the MSRT topology, let us first start from 1D MSRT bypass rings. A 1D MSRT bypass ring originates from a 1D SRT ring by eliminating every other bypass link. In, 1D MSRT ($L = 2$; $l_1 = 2, l_2 = 4$) is a truncated 1D SRT ($L = 2$; $l_1 = 2, l_2 = 4$). $L = 2$ is the maximum node level. This means that two types of bypass links exist, i.e. l_1 and l_2 links. Then, $l_1 = 2$ and $l_2 = 4$ indicates the short and long bypass links spanning over $2^{l_1} = 4$ and $2^{l_2} = 16$ hops respectively. Figure 3.6 shows the similarity and difference of SRT and MSRT. We extend 1D MSRT bypass rings to 3D MSRT networks. To maintain a suitable node degree, we add two types of bypass links in x-axis and y-axis and then form a 2D expansion network in xy-plane.

To study the advantages of MSRT networks, we compared the MSRT with other networks' performance metric in Table 3.6. For calculating the average distances of 3D MSRT, we first select a l_1-level bypass node as the reference node and then a l_2-level bypass node so two results are present in the left and right columns respectively.

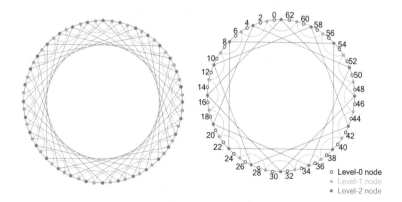

Fig. 3.6. 1D SRT ring and corresponding MSRT ring.

Table 3.6.   Analytical comparison between MSRT and other topologies.

Topology	Bypass networks	Dimensions	Node degree	Diameter (hop)	Bisection width (×1024) links)	Average distance (hop)
3D Torus	N/A	32 × 32 × 16	6	40	1	20.001
4D Torus	N/A	16 × 16 × 8 × 8	8	24	2	12.001
6D Torus	N/A	8 × 8 × 4 × 4 × 4 × 4	12	16	4	8.000
3D SRT	$L = 1$; $l_1 = 4$	32 × 32 × 16	10	24	9	12.938
2D SRT	$l_{max} = 6$	128 × 128	8	13	1.625	8.722
3D MSRT	$L = 2$; $l_1 = 2$, $l_2 = 4$	32 × 32 × 16	8	16	2.5	9.239 9.413

3.2.2. *Interconnect Technology*

Along with the network topology goes the interconnect technology that enables the fancy designed network to achieve its maximum performance. Unlike the network topology that settles down to several typical schemes, the interconnect technology shows a great diversity ranging from the commodity Gigabit Ethernet all the way to the specially designed proprietary one (Fig. 3.7).

3.3. Instruction and Data Streams

Based on the nature of the instruction and data streams (Fig. 3.8), parallel computers can be made as SISD, SIMD, MISD, MIMD where I stands for Instruction, D for Data, S for Single, and M for Multiple.

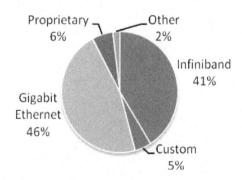

Fig. 3.7.   Networks for Top500 supercomputers by system count in June 2011.

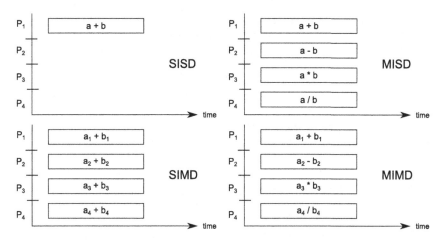

Fig. 3.8. Illustration of four instruction and data streams.

An example of SISD is the conventional single processor workstation. SIMD is the single instruction multiple data model. It is not very convenient to be used for wide range of problems, but is reasonably easy to build. MISD is quite rare. MIMD is very popular and appears to have become the model of choice, due to its wideness of functionality.

3.4. Processor-Memory Connectivity

In a workstation, one has no choice but to connect the single memory unit to the single CPU, but for a parallel computer, given several processors and several memory units, how to connect them to deliver efficiency is a big problem. Typical ones are, as shown schematically by Figs. 3.9 and 3.10, are:

➤ Distributed-memory.
➤ Shared-memory.
➤ Shared-distributed-memory.
➤ Distributed–shared-memory.

3.5. IO Subsystems

High-performance input-output subsystem is the second most essential component of a supercomputer and its importance grows even more

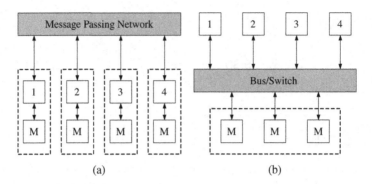

Fig. 3.9.   (a) Distributed-memory model; (b): Shared-memory model.

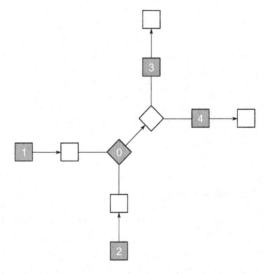

Fig. 3.10.   A shared-distributed-memory configuration.

apparent with the latest introduction of Web 2.0 services such as video stream, e-commerce, and web serving.

Two popular files systems are Lustre and IBM's GPFS.

Lustre is a massively parallel distributed file system, generally used for large scale cluster computing and it provides a high performance file system for clusters of tens of thousands of nodes with petabytes of storage capacity. It serves computer clusters ranging from small workgroup to large-scale, multi-site clusters. More than half of world's top supercomputers use Lustre file systems.

Lustre file systems can support tens of thousands of client systems, tens of petabytes (PBs) of storage and hundreds of gigabytes per second (GB/s) of I/O throughput. Due to Lustre's high scalability, internet service providers, financial institutions, and the oil and gas industry deploy Lustre file systems in their data centers.

IBM's **GPFS** is a high-performance shared-disk clustered file system, adopted by some of the world's top supercomputers such as ASC purple supercomputer which is composed of more than 12,000 processors and has 2 petabytes of total disk storage spanning more than 11,000 disks.

GPFS provides tools for management and administration of the GPFS cluster and allows for shared access to file systems from remote GPFS clusters.

3.6. System Convergence

Generally, at least three parameters are necessary to quantify the quality of a parallel system. They are:

(1) Single-node performance.
(2) Inter-node communication bandwidth.
(3) I/O rate.

The higher these parameters are, the better would be the parallel system. On the other hand, it is the proper balance of these three parameters that guarantees a cost-effective system. For example, a narrow inter-node width will slow down communication and make many applications unscalable. A low I/O rate will keep nodes waiting for data and slow overall performance. A slow node will make the overall system slow.

3.7. Design Considerations

Processors: Advanced pipelining, instruction-level parallelism, reduction of branch penalties with dynamic hardware prediction and scheduling.

Networks: Inter-node networking topologies and networking controllers.

Processor-memory connectivity: (1) Centralized shared-memory, (2) distributed-memory, (3) distributed–shared-memory, and (4) virtual shared-memory. Naturally, manipulating data spreading on such memory system is a daunting undertaking and is arguably the most significant challenge of all aspects of parallel computing. Data caching, reduction of caching misses and the penalty, design of memory hierarchies, and virtual memory are all subtle issues for large systems.

Storage Systems: Types, reliability, availability, and performance of storage devices including buses-connected storage, storage area network, raid.

In this "4D" hardware parameter space, one can spot many points, each representing a particular architecture of the parallel computer systems. Parallel computing is an application-driven technique, where unreasonable combination of computer architecture parameters are eliminated through selection quickly, leaving us, fortunately, a small number of useful cases. To name a few: distributed-memory MIMD (IBM BlueGene systems on 3D or 4D torus network, Cray XT and XE systems on 3D torus network, China's National Defense University's Tianhe-1A on fat tree network, and Japan's RIKEN K computer's 6D torus network), shared-memory MIMD (Cray and IBM mainframes), and many distributed shared-memory systems.

It appears that distributed-memory MIMD architectures represent optimal configurations.

Of course, we have not mentioned an important class of "parallel computers", i.e. the Beowulf clusters by hooking up commercially available nodes on commercially available network. The nodes are made of the off-the-shelf Intel or AMD X86 processors widely marketed to the consumer space in the number of billions. Similarly, such nodes are connected on Ethernet networks, either with Gigabit bandwidth in the 2000s or 10-Gigabit bandwidth in the 2010s, or Myrinet or Infiniband networks with much lower latencies. Such Beowulf clusters provide high peak performance and are inexpensive to purchase or easy to make in house. As a result, they spread like fire for more than a decade since the 1990s. Such clusters are not without problems. First, the power efficiency (defined as the number of sustained FLOPS per KW consumed) is low. Second, scaling such system to contain hundreds of thousands of cores is a mission impossible. Third, hardware reliability suffers badly as system size increases. Fourth, monetary cost can grow nonlinearly because of the needs of additional networking. In summary, Beowulf clusters have served the computational science community profitably for a generation of researchers and some new innovation in supercomputer architectures must be introduced to win the next battle.

CHAPTER 4

SOFTWARE SYSTEMS

Software system for a parallel computer is a monster as it contains the node software that is typically installed on serial computers, system communication protocols and their implementation, libraries for basic parallel functions such as collective operations, parallel debuggers, and performance analyzers.

4.1. Node Software

The basic entity of operation is a processing node and this processing node is similar to a self-contained serial computer. Thus, we must install software on each node. Node software includes node operating system, compilers, system and applications libraries, debuggers, and profilers.

4.1.1. *Operating systems*

Essentially, augmenting any operating system for serial computers to provide data transfers among individual processing units can constitute a parallel computing environment. Thus, regular serial operating systems (OS) such as UNIX, Linux, Microsoft Windows, and Mac OS are candidates to snowball for a parallel OS. Designing parallel algorithms for large-scale applications requires full understanding of the basics of serial computing. UNIX is well developed and widely adopted operating system for scientific computing communities. As evident in the picture (Fig. 4.1) copied from Wikipedia, UNIX has gone through more than four decades of development and reproduction.

Operating systems such as Linux or UNIX are widely used as the node OS for most parallel computers with their vendor flavors. It is notimpossible to see parallel computers with Windows installed as their node OS, however, It is rare to see Mac OS as the node OS in spite of their huge commercial success.

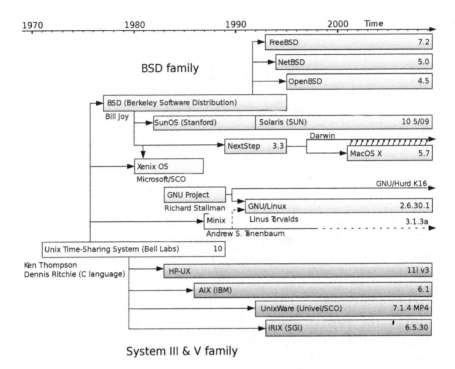

Fig. 4.1.   The evolutionary relationship of several UNIX systems (Wikipedia).

Linux is a predominantly node OS, and one of the most successful free and open source software packages. Linux, with its many different variations, is installed on a wide variety of hardware devices including mobile phones, tablet computers, routers and video game consoles, laptop and desktop computers, mainframes, and most relevantly the nodes of parallel computers and it runs on the nodes of most, if not all, Top500 supercomputers. A distribution intended to run on supercomputer nodes usually omits many functions such as graphical environment from the standard release and instead include other software to enable communications among such nodes.

4.1.2. *Compilers and libraries*

We limit our discussion on compilers that transform source code written in high-level programming languages into low-level executable code for parallel computers' nodes, i.e. serial elements. All compilers always perform the following operations when necessary: Lexical analysis, preprocessing, parsing, semantic analysis, code optimization, and code generation. Most commonly, a node-level executable program capable of communicating with other executable programs must be generated by a serial language compiler. The popular high-level languages are FORTRAN, C, and C++. Many other serial computer programming languages can also be conveniently introduced to enable parallel computing through language binding.

Parallel compilers, still highly experimental, are much more complex and will be discussed later.

A computer **library** is a collection of resources in form of subroutines, classes, and values, etc., just like the conventional brick-and-mortar libraries storing books, newspapers, and magazines. Computer libraries contain common code and data for independent programs to share code and data to enhance modularity. Some executable programs are both standalone programs and libraries, but most libraries are not executable. Library objects allow programs to *link* to each other through the process called *linking* done by a linker.

4.1.3. *Profilers*

A computer profiler is a computer program that performs dynamic program analysis of the usages of memory, of particular instructions, or frequency and duration of function calls. It became a part of either the program source code or its binary executable form to collect performance data at run-time.

The most common use of profiling information is to aid program optimization. It is a powerful tool for serial programs. For parallel programs, a profiler is much more difficult to define let alone to construct. Like parallel compilers and parallel debuggers, parallel profilers are still work in progress.

4.2. Programming Models

Most early parallel computing circumstances were computer specific, with poor portability and scalability. Parallel programming was highly dependent on system architecture. Great efforts were made to bridge the gap between parallel programming and computing circumstances. Some standards were promoted and some software was developed.

Observation: A regular sequential programming language (like Fortran, C, or C++, etc.) and four communication statements (send, recv, myid, and numnodes) are necessary and sufficient to form a parallel computing language.

send: One processor sends a message to the network. The sender does not need to specify the receiver but it does need to designate a "name" to the message.

recv: One processor receives a message from the network. The receiver does not have the attributes of the sender but it does know the name of the message for retrieval.

myid: An integer between 0 and $P-1$ identifying a processor. myid is always unique within a partition.

numnodes: An integer which shows the total number of nodes in the system.

This is the so-called single-sided message passing, which is popular in most distributed-memory supercomputers.

It is important to note that the Network Buffer, as labeled, in fact, does not exist as an independent entity and is only a temporary storage. It is created either in the senders RAM or in the receivers RAM and is dependent on the readiness of the message routing information. For example, if a message's destination is known but the exact location is not known at the destination, the message will be copied to the receivers RAM for easier transmission.

4.2.1. *Message passing*

There are three types of communication in parallel computers. They are synchronous, asynchronous, and interrupt communication.

Synchronous Communication: In synchronous communication, the sender will not proceed to its next task until the receiver retrieves the message from the network. This is analogous to hand delivering a message to someone, which can be quite time consuming (Fig. 4.2).

Asynchronous Communication: During asynchronous communication, the sender will proceed to the next task whether the receiver retrieves the message from the network or not. This is similar to mailing a letter. Once the letter is placed in the post office, the sender is free to resume his or her activities. There is no protection for the message in the buffer (Fig. 4.3).

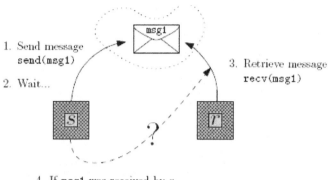

Fig. 4.2.   An illustration of synchronous communication.

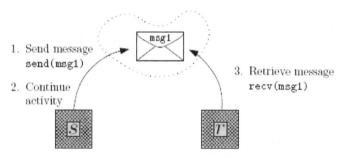

Fig. 4.3.   An illustration of asynchronous communication.

Asynchronous message passing example: The sender issues a message, then continues with its tasks regardless of the receivers' response in treating the message (Fig. 4.4). While the receiver has four options with regard to the message issued already by the sender, this message now stays somewhere called the buffer:

(1) Receiver waits for the message.
(2) Receiver takes the message only after it has finished its current task, either computing or communicating with another processor.
(3) Receiver ignores the message.
(4) Receiver merges this message to another message still residing in the buffer.

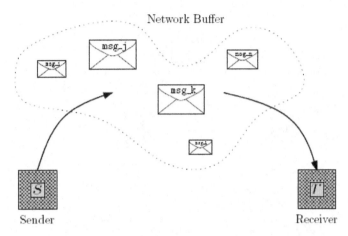

Fig. 4.4. Sending and receiving with the network buffer.

Interrupt: The receiver interrupts the sender's current activity in order to pull messages from the sender. This is analogous to the infamous 1990s telemarketer's calls at dinnertime in US. The sender issues a short message to interrupt the current execution stream of the receiver. The receiver becomes ready to receive a longer message from the sender. After an appropriate delay (for the interrupt to return the operation pointer to the messaging process), the sender pushes through the message to the right location of the receiver's memory, without any delay (Fig. 4.5).

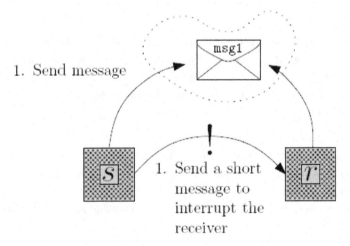

Fig. 4.5. An illustration of interrupt communication.

Communication Patterns: There are nine different communication patterns:

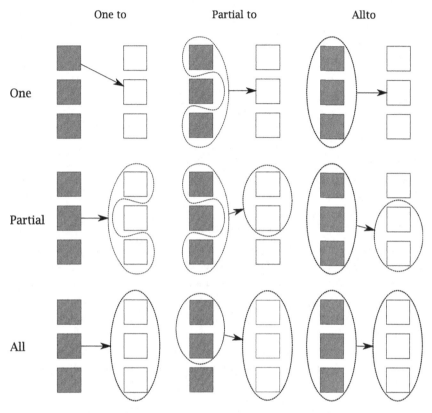

Four Modes of MPI Message Passing:

Standard:

Assumes nothing on the matching send.
Assumes nothing on the matching receives.
Messages are usually not buffered for performance gain.

Buffered:

When buffering, messages stay in sender's or receiver's memory or both.
Moving messages to the destination from buffer as soon as destination is known and is ready.

Synchronous:

A request is sent to receiver and, when acknowledged, pushes the message to the receiver.
When messages arrive, both sender and receiver move on.

Ready:

Sender knows, before sending, that the matching receiver has already been posted to receive.

4.2.2. *Shared-memory*

Shared-memory computer (Fig. 4.6) is another large category of parallel computers. As the name indicates, all or parts of the memory space are shared among all the processors. That is, the memory can be simultaneously accessed directly by multiple processors. Under this situation, communications among processors are carried out, implicitly, by accessing the same memory space from different processors, at different times. This scheme permits the developer an easier task to write parallel program, whereby at least the data sharing is made much more convenient. Serious problems such as system scalability restricted by limited data sharing are still unsolved. As we will discuss later, multiple techniques were introduced to mitigate such challenges.

Although all the memory can be shared among processors without extra difficulty, in practical implementation, memory space on a single node has

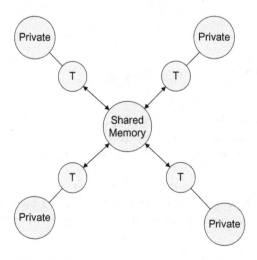

Fig. 4.6.   Schematic diagram for a typical shared-memory computer.

always been partitioned into shared and private part, at least logically, in order to provide flexibility in program development. Shared data is accessible by all but private data can only be accessed by the one who owns it.

4.3. Parallel Debuggers

A parallel debugger is no different from a serial debugger. dbx is one popular example of serial debugger. The dbx is a utility for source-level debugging and execution of programs written in C and FORTRAN.

Most parallel debuggers are built around dbx with additional functions for handling parallelism, for example, to report variable addresses and contents in different processors. There are many variations to add convenience by using graphic interface.

A typical debugger, interactive parallel debugger (IPD), is a complete symbolic, source-level debugger for parallel programs. Beyond the standard operations that facilitate the debugging of serial programs, IPD offers custom features that facilitate debugging parallel programs. IPD lets you debug parallel programs written in C, FORTRAN, and assembly language. IPD consists of a set of debugging commands, for which help is available from within IPD.

4.4. Parallel Profilers

Similar to profilers for serial computing, parallel profilers provide tools for analyzing parallel program execution including serial performances, communication performance, and event traced performance of application programs.

Many parallel profilers are in circulation and none appear dominating. A profiler, ParaGraph, introduced in the early 2000s, is likely to have been forgotten but its design goals and expected functionalities are still not obsolete.

ParaGraph is a graphical display system for visualizing the behavior and performance of parallel programs on message-passing multiprocessor architectures. This is an ancient utility but its philosophy is still update-to-date. It takes as input execution profile data provided by the Portable Instrumented Communication Library (PICL) developed at Oak Ridge National Laboratory. ParaGraph is based on the X Window System, and thus runs on a wide variety of graphical workstations. The user interface for ParaGraph is menu-oriented, with most user input provided by mouse clicks, although for some features, keyboard input is also supported. The execution of ParaGraph is event driven, including both user-generated X Window events and trace events in the data file produced by PICL. Thus, ParaGraph provides a dynamic depiction of the parallel program while also providing responsive interaction with the user. Menu selections determine the execution behavior of ParaGraph both statically and dynamically. As a further aid to the user, ParaGraph preprocesses the input trace file to determine relevant parameters automatically before the graphical simulation begins.

CHAPTER 5

DESIGN OF ALGORITHMS

Quality algorithms, in general, must be (1) accurate, (2) efficient, (3) stable, (4) portable, and (5) maintainable.

For parallel algorithms, the quality metric "efficient" must be expanded to include high parallel efficiency and scalability.

To gain high parallel efficiency, two factors must be considered:

(1) Communication costs;
(2) Load imbalance costs.

For scalability, there are two distinct types: Strong scaling and weak scaling. A strong scaling algorithm is such that it is scalable for solving a fixed total problem size with a varying number of processors. Conversely, a weak scaling algorithm is such that it is scalable for solving a fixed problem size per processor with a varying number of processors.

Minimizing these two costs simultaneously is essential but difficult for optimal algorithm design. For many applications, these two factors compete to waste cycles and they have many dependencies. For example, one parameter that one can always adjust for performance — granularity. Usually, granularity depends on the number of processors, and the smaller the granularity, the smaller load imbalance and the bigger the communication.

More specifically, efficient parallel algorithms must satisfy many or all of the following conditions:

(1) Communication to computation ratio is minimal.
(2) Load imbalance is minimal.
(3) Sequential bottleneck is minimal.
(4) Non-parallel access to storage is minimal.

Achieving all is likely to ensure the high parallel efficiency of a parallel algorithm. How do we achieve it is described in the following sections.

5.1. Algorithm Models

The paradigm of divide and conquer, tried and true in sequential algorithm design is still the best foundation for parallel algorithms. The only step that needs consideration is "combine". Thus the key steps are:

(1) Problem decompositions (data, control, data + control).
(2) Process scheduling.
(3) Communication handling (interconnect topology, size and number of messages).
(4) Load Balance.
(5) Synchronization.
(6) Performance analysis and algorithm improvement.

The following basic models are widely utilized for design of efficient parallel algorithms:

(1) Master-slave.
(2) Domain decomposition.
(3) Control decomposition.
(4) Data parallel.
(5) Single program multiple data. (SPMD)
(6) Virtual-shared-memory model.

Of course, other parallel programming models are also possible. For a large application package, it is common to utilize two or more programming models.

5.1.1. *Master-slave*

A master-slave model is the simplest parallel programming paradigm with an exception of the embarrassingly parallel model. In the master-slave model, a master node controls the operations of the slave nodes in the system. This model can be applied to many types of applications and is also popular among the initial practitioners of parallel computing. But it has serious drawbacks for sophisticated applications in which a single "master" or even a series of strategically distributed "masters" are unable to handle the demands of the "slaves" smoothly to minimize their idling. Thus, this model is usually, and should be, avoided for serious and larger scale applications.

Figure 5.1 illustrates the Master-slave model.

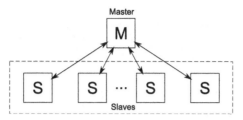

Fig. 5.1. A sketch of the Master-slave model in which communications occur only between master and individual slaves while communications never occur among slaves.

5.1.2. *Domain decomposition*

Domain decomposition, invented much earlier in numerical analysis for particularly numerical solutions of PDEs, is defined as a method of splitting a large problem domain into smaller ones. It is essentially a "divide and conquer" technique for arriving at the solution of an operator equation posed on a domain from the solution of related operator equations posed on sub-domains. This borrowed term for parallel computing has several definitions:

Definition I: The domain associated with a problem is divided into parts, or grains, one grain per processor of the parallel computer. In the case of domains with a metric (e.g., typical spatial domains) we use domain decomposition in a more specific sense to denote the division of space into locally connected grains.[1]

Definition II: In domain decomposition, the input data (the domain) is partitioned and assigned to different processors. Figure 5.2 illustrates the data decomposition model.

5.1.3. *Control decomposition*

In control decomposition, the tasks or instruction streams rather than data are partitioned and assigned to different processors. Control decomposition approach is not as uncommon as it appears for many non-mathematical applications.

Figure 5.3 illustrates control decomposition.

---

[1] Fox Johnson Lyzenga Otto Salmon Walker.

Domain

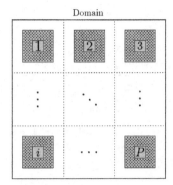

Fig. 5.2.   A sketch of domain decomposition model in which the computational domain is decomposed to sub-domains (which are then assigned to computing resources). This model is popular for solving problems defined on obvious physical spaces, e.g., many PDEs. It is almost the standard approach for decomposing problems with locality.

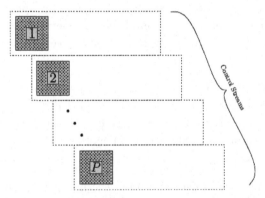

Fig. 5.3.   In this model, the computational domain is decomposed into multiple sub-domains each of which is assigned to a process. This model is useful for implementing SIMD and MIMD, and it is popular for solving PDEs. This domain decomposition approach is particularly effective for solving problems with locality.

5.1.4. *Virtual-shared-memory*

Shared-memory model has the greatest advantage of accessing data without explicit message passing because the entire (global) data space is shared among participating processors. It has a serious drawback in achieving scalability for larger systems with 16 processors or a similar magic upper limit due to the greatest inability to share data harmoniously, among others. Wise attempts came to the rescue: Virtual shared memory model in which data are shared virtually by aid of software. To the applications developers or the struggling programmers, there is an ocean of memory to which all participating processors, regardless of their physical connections to the memory banks, can read and write data conveniently without any explicit message-passing statements. The housekeeping tasks such as strategically addressing the read and write is the business of the demonic agents.

For better or for worse, the programmer no longer needs to explicitly manage message passing and, of course, no longer controls and exploits message passing for better efficiency. The gain of removing explicit message passing hassle is matched by loss of program efficiency. As in most programming trade-offs, one would gain some and lose some.

The fact that this model causes major overheads of resources makes it a wishful academic exercise although it is likely a future model of choice for data sharing in parallel computing.

Figure 5.4 illustrates the virtual-shared-memory model.

5.1.5. *Comparison of programming models*

(1) Explicit vs. implicit.
(2) Virtual-shared-memory vs. message passing.
(3) Data parallel vs. control parallel.
(4) Master-slave vs. the rest.

5.1.6. *Parallel algorithmic issues*

An important issue in creating efficient algorithms for parallel computers (or for any computer in general) is the balancing of:

(1) The programmer and computer efforts (see Fig. 5.5).
(2) Code clarity and code efficiency.

The amount of effort devoted by a programmer is always correlated by the efficiency and program clarity. The more you work on the algorithms, the more the computer will deliver. The broad spectrum of approaches for creating a parallel algorithm, ranging from brute force to state-of-the-art software engineering, allows flexibility in the amount of effort devoted by a programmer.

For short-term jobs, ignoring efficiency for quick completion of the programs is the usual practice. For grand challenges that may run for years for a set of parameters, careful designs of algorithms and implementation are required. Algorithms that are efficient may not be clean and algorithms

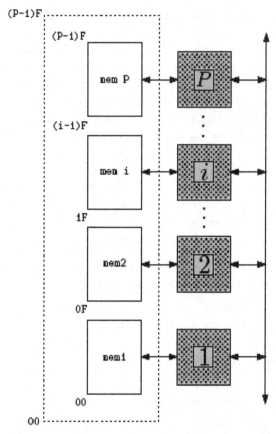

Fig. 5.4. In virtual-shared-memory model, each process "owns" an entire memory space virtually. In reality, each process only owns a portion of the entire memory space while the operating system virtualizes the space and facilities data transfers of data when necessary.

that are clean may not be efficient. Which is more important? There is no universal rule. It is problem-dependent. The desirable programs must be both clean and efficient.

The following are the main steps involved in designing a parallel algorithm:

(1) Decomposing a problem in similar sub-problems.
(2) Scheduling the decomposed sub-problems on participating processors.
(3) Communicating the necessary information between the processors so that each processor has sufficient information for progressing.

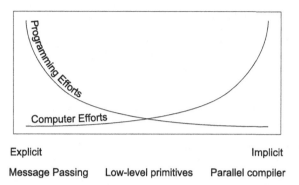

Fig. 5.5. This figure illustrates the relationship between programmer time and computer time, and how it is related to the different ways of handling communication.

(4) Collecting results from each processor to form the final result to the original problem.

There are key tasks for designing an efficient parallel algorithm:

(1) Construction of serial algorithm for each processing unit.
(2) Handlings of message passing among processing units.
(3) Balancing loads among processing units.

A parallel-efficient algorithm should always have minimal overheads of load imbalance and communication. These two factors usually tend to corrupt, making algorithms inefficient; to minimize them individually and simultaneously is the highest design objective that can be costly, for some applications, to achieve. For some problems with intrinsic local interactions such as the hyperbolic equations and short-ranged molecular dynamics, to gain minimal communication, one must minimize the aspect ratio of the sub-domains allocated to each processor (assuming more than one sub-domain can be given to each processor), which can happen only when each processor has one very round-shaped sub-domain. This choice of sub-domain size, however, increases the variance in loads on all processors and creates extreme imbalance of loads. On the other hand, if the sub-domains are made small in size, load balance may be achieved easily, but communication soars. Conventional wisdom tells us to choose a sub-domain with medium size.

Minimizing the overhead due to load imbalance and communication by choosing a proper granularity is the main goal in designing parallel algorithms. A good algorithm must be robust, which includes reliability, flexibility, and versatility.

5.1.7. *Levels of algorithmic complication*

Being a new form of art, parallel computing could be extremely simple or extremely difficult depending on applications. So, application problems are classified into three classes as shown in Fig. 5.6.

Embarrassingly parallel

The lowest class is the embarrassingly parallel problem in which very little communication is needed, if ever; it might only occur once at the

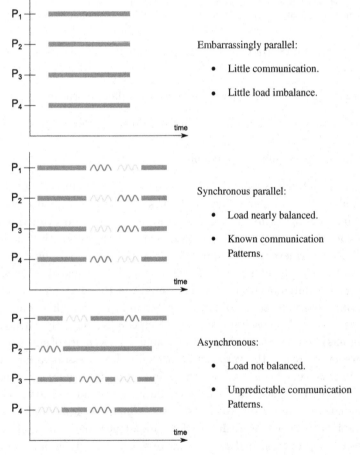

Fig. 5.6.   An illustration of parallel complexity.

beginning to kick off the parallel executions on all processors or at the end to collect results from all the processors. Furthermore, load balance is automatically preserved. Algorithms for this type of problems are always close to 100% parallel efficient and scalable. Examples of embarrassingly parallel algorithms include all numerical integration and most of the master-slave type problems.

Synchronous parallel

The next class is the synchronized parallel problem in which communications are needed to synchronize the executions of a fixed cluster of processors in the middle of a run. Load is also almost balanced if the synchronization is done properly. Algorithms for this type of problems may achieve very high parallel efficiencies and they are quasi-scalable with reasonably large scaling zones.

Most problems involving local interactions such as the hyperbolic PDEs or classical short-ranged molecular dynamics are of this type.

Asynchronous parallel

The most difficult class is the asynchronous parallel problem in which one can hardly formulate any communication patterns and most communications involving the entire system of processors. The load is also non-deterministic and thus rarely possible to balance statically. Algorithms for this type of problem always have small parallel efficiency and are rarely scalable.

It is easy to understand why they are not scalable. For a fixed-size problem, the computational load is nearly fixed. When the number of processors increase, the total communication increases, leading to decrease in parallel efficiency. This type of problem is poisonous to parallel computing. The unfortunate truth is that most physically-interesting problems including linear algebra problems such as LU decomposition and matrix multiplications, elliptical PDEs, CMD with long-ranged interactions, quantum molecular dynamics, and plasma simulations belong to this class.

Does this mean parallel computing is useless for scientific problems? The answer is 'No'. The reason is that, in reality, when increasing the number of processors, one can always study relatively larger problems. A parallel efficiency of 50% does not pose a big challenge, but it can always make a run faster (sometimes by orders of magnitude) than a serial run. In fact, serial computing is indeed a dead end for scientific computing.

5.2. Examples of Collective Operations

To broadcast a string of numbers (or other data) to a 1D array of processors and to compute a global sum over numbers scattered on all processors are both very simple but typical examples of parallel computing.

5.2.1. *Broadcast*

Suppose processor 0 possesses N floating-point numbers $\{X_{N-1}, \ldots, X_1, X_0\}$ that need to be broadcast to the other $P-1$ processors in the system. The best way to do this (Figs. 5.7 and 5.8) is to let P_0 send out X_0 to F_1, which then sends X_0 to P_2 (keeping a copy for itself) while P_2 sends the next number, X_1, to P_1. At the next step, while P_0 sends out X_2 to P_1, P_1 sends out X_1 to P_2, etc., in a pipeline fashion. This is called the wormhole communication.

Suppose the time needed to send one number from a processor to its neighbor is T_{comm}. Also, if a processor starts to send a number to its neighbor at time T_1 and the neighbor starts to ship out this number at time T_2, we define a parameter called:

$$T_{\text{startup}} = T_2 - T_1. \tag{5.1}$$

Typically, $T_{\text{startup}} \geq T_{\text{comm}}$. Therefore, the total time needed to broadcast all N numbers to all P processors is given by:

$$T_{\text{comm}} = NT_{\text{comm}} + (P-2)T_{\text{startup}}. \tag{5.2}$$

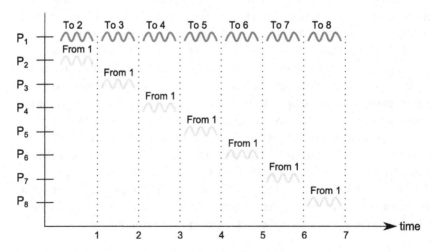

Fig. 5.7. Broadcast model 1.

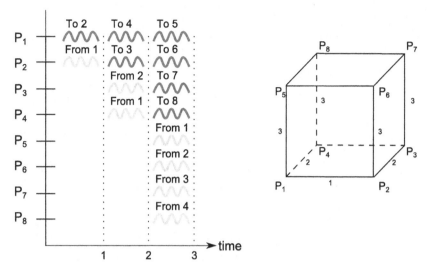

Fig. 5.8. Broadcast model 2.

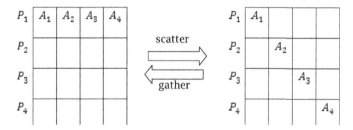

Fig. 5.9. Illustration of gather and scatter.

5.2.2. *Gather and scatter*

In this scheme, as shown in Fig. 5.9,

(1) The processors on the two ends of the arrows labeled 1 perform a message exchange and summation.
(2) The processors on the two ends of the arrows labeled 2 perform a message exchange and summation.
(3) The processors on the two ends of the arrows labeled 3 perform a message exchange and summation.

At the end of the third step, all processors have the global sum.

5.2.3. *Allgather*

We explain the global summation for a 3D hypercube. Processor P_t contains a number X_1 for $i = 1, 2, \ldots, 7$. We do the summation in three steps.

Step 1: Processors P_0 and P_1 swap contents and then add them up. So do the pairs of processors P_2 and P_3, P_4 and P_5, as well as P_6 and P_7. All these pairs swap and add corresponding data in parallel. At the end of this parallel process, each processor has its partner's content added to its own, e.g. P_0 now has $X_0 + X_1$.

Step 2: P_0 exchanges its new content with P_2 and performs addition (other pairs follow this pattern). At the end of this stage, P_0 has $X_0 + X_1 + X_2 + X_3$.

Step 3: Finally, P_0 exchanges its new content with P_4 and performs addition (again, other pairs follow the pattern). At the end of this stage, P_0 has $X_0 + X_1 + X_2 + X_3 + X_4 + X_5 + X_6 + X_7$.

In fact, after $\log_2 8 - 3$ stages, every processor has a copy of the global sum. In general for a system containing P processors, the total time needed for global summation is $O(\log_2 P)$.

5.3. Mapping Tasks to Processors

Figure 5.11 charts the pathway of mapping an application to the topology of the computer. Several mapping functions as shown in Fig. 5.10 can be implanted to realize the following mappings:

(1) Linear Mapping.
(2) 2D Mapping.
(3) 3D Mapping.
(4) Random Mapping: flying processors all over the communicator.
(5) Overlap Mapping: convenient for communication.
(6) Any combination of the above.

Two benefits are performance gains and code readability.

In Table 5.1, we illustrate the mapping of a 1D application to 2D computer architecture.

This helps create a balanced distribution of processes per coordinate dimension, depending on the processes in the group to be balanced and optional constraints specified by the user. One possible use is to partition all processes into an n-dimensional topology. For example,

Call	Dims returned
MPI_Dims_create(6,2,dims)	3,2
MPI_Dims_create(7,2,dims)	7,1 or 1,7
MPI_Dims_create(6,3,dims)	3,2,1 or 2,3,1 or 1,2,3
MPI_Dims_create(7,3,dims)	no answer

```
MPI_Cart_get(/* Obtaining more details on a Cartesian communicator
*/
      IN comm,
      IN maxdims, /* length of vector dims, periods, coords */
          number_of_processes_in_each_dim,
          perodis_in_each_dim,
          cords_of_calling_process_in_cart_structure
)

/* Cartesian translator function: rank <===> Cartesian */

MPI_Cart_rank(/* rank => Cartesian */
      IN comm,
      IN coords,
          rank
)

MPI_Cart_coords(/* rank => Cartesian */
      IN comm,
      IN rank,
      IN maxdims,
          coords /* coords for specific processes */
)

/* Cartesian partition (collective call) */

MPI_cart_sub(
      IN comm,
      IN remain_dims, /* logical variable TRUE/FALSE */
          new_comm
)
```

Fig. 5.10.   Some Cartesian inquire functions (local calls).

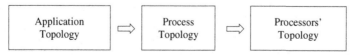

Fig. 5.11.   This is a "Process-to-Processor" mapping.

Table 5.1.   A 1D application
is mapped to a 2D computer
architecture.

0	1	2
(0,0)	(0,1)	(0,2)
3	4	5
(1,0)	(1,1)	(1,2)
6	7	8
(2,0)	(2,1)	(2,2)

0	1	2
(0,0)	(0,1)	(0,2)
3	4	5
(1,0)	(1,1)	(1,2)
6	7	8
(2,0)	(2,1)	(2,2)

```
MPI_Cart_create(/* Creating a new communicator */
                /* (row-majored ranking)       */
        old_comm,
        number_of_dimensions_of_cart,
        array_specifying_meshs_in_each_dim,
        logical_array_specifying_periods, /* TRUE/FALSE */
        reoreder_ranks?,
        comm_cart /* new communicator */
)

MPI_Dims_create(
        number_nodes_in_grid,
        number_cartesian_dims,
        array_specifying_meshs_in_each_dim
)
```

Fig. 5.12.   MPI functions.

The functions listed in Fig. 5.12 are useful when embedding a lower
dimensional Cartesian grid into a bigger one, with each sub-grid forming a
sub-communicator.

If the `comm` $= 2 \times 3 \times 4$ 24 processes we have
if `remain_dimms` $=$ (`true`, `false`, `true`)
`new_comm` $= 2 \times 4$ (8 processes)

All processors in parallel computers are always connected directly or indirectly. The connectivity among processors measured in latency, bandwidth and others are non-uniform and they depend on the relative or even absolute locations of the processor in the supercomputer. Thus, the way to assign computation modules or subtasks on nodes may impact the communication costs and the total running. To optimize this part of the running time, we need to utilize the task mapping.

According to Bokhari in his *On the Mapping Problem*,[2] the mapping and the mapping problem can be defined as follows:

Suppose a problem made up of several modules that execute in parallel is to be solved on an incompletely connected array. When assigning modules to processors, pairs of modules that communicate with each other should be placed, as far as possible, on processors that are directly connected. We call the assignment of modules to processors a mapping and the problem of maximizing the number of pairs of communicating modules that fall on pairs of directly connected processors the mapping problem.

The application and the parallel computer are represented as static graphs G_A and G_P, respectively. $G_A = (V_A; E_A)$ is a graph where V_A represents tasks and E_A means communication requests among them with weights being the communication loads. $G_P = (V_P; E_P)$ is a graph where V_P represents processors and E_P means the links among processors with weights being the per unit communication cost.

A mapping problem can then be defined as finding a mapping: $V_A \rightarrow V_P$ to minimize the objective function value that associates with each mapping.

5.3.1. *Supply matrix*

Imitating business terminology, we introduce the supply-and-demand concept for task mapping. A supply matrix of a supercomputer network

---

[2]S. H. Bokhari, On the mapping problem, *IEEE Transactions on Computers.* **30**(3) (1981) 207–214.

is a measure of the connectivity of the supercomputer's interconnection networks. For example, the matrix element S_{ij} measures the network latency for initiating data transmission between processors i and j.

Hop matrix

Hop matrix measures the distance among all nodes in terms of minimal number of hops between all pairs of nodes. Such hop matrix is a decent approximation of the latency matrix measuring the delays or distances messages traversing pairs of nodes. We must be aware that this approximation can be rough as collisions, among others, can easily occur and messages may not always travel on shortest-distance paths.

Latency matrix

The latency matrix (Fig. 5.13) is measured from the communication time for sending a 0 byte message from one node to all other nodes that forms a matrix.

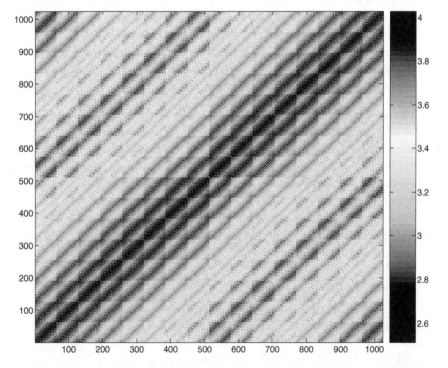

Fig. 5.13.   MPI latency of a 0 byte packet on a 8*8*16 BG/L system.

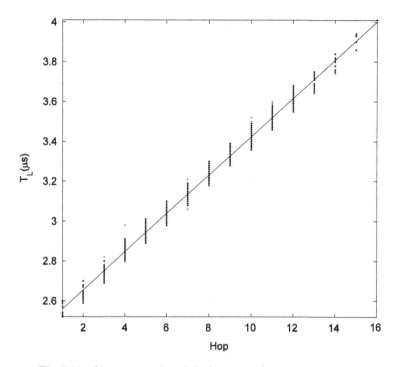

Fig. 5.14. Linear regression of the latency with respect to the hop.

As shown in Fig. 5.14, linear regression analysis of the latency with respect to the hop shows that the hops match the latencies remarkably nicely albeit many outliers that may mislead the optimization, i.e. the task mapping. Most of the differences result from the torus in Z-dimension of the BG/L system.

5.3.2. *Demand matrix*

A demand matrix is a metric measuring the amount of data that need to be transmitted between all pairs of subtasks. Each matrix element D_{ij} measures the total data during a period of time, in bytes, for example, from subtask i to subtask j. Usually, the period of time is chosen as the lifetime of an application. This is for static mapping and it is quite restrictive and limiting. We may discretize the period of time to smaller intervals measuring the characteristic time scales of message passing. This can be the starting point of dynamic task mapping whose research, let alone practical software, is still at its early stages.

5.3.3. *Review of mapping models*

The general idea of the static mapping is to build a model where the total communication time can be calculated based on the mapping. After the model is built, the problem turns into an optimization problem and thus can be solved by existing optimization techniques.

By considering different aspects and complexity, here we introduce some of the important mapping models.

Model by Bokhari

In 1981, Bokhari proposed a model that maps n tasks to n processors to find the minimum communication cost independent of computation costs which can be represented as:

$$\min \sum_{x \in V_p, y \in V_p} G_p(x, y) \cdot G_a(f_m(x), f_m(y)). \tag{5.3}$$

This model is used in the analysis of algorithms for SIMD architectures. Bokhari points out that this model can be transformed to a graph isomorphism problem. However, it cannot reflect the amount differences among inter-module communications and inter processor likes.

Model by Heiss and Dormanns

In 1996, Heiss and Dormanns formulated the mapping problem as to find a mapping $T \rightarrow P$

$$\min CC = \sum_{(i,j) \in E^T} a(i, j) \cdot d(\pi(i), \pi(j)), \tag{5.4}$$

where $a(i, j)$ is the amount of data needed to be exchanged and $d(\pi(i), \pi(j))$ represents the length of shortest path between i and j. Heiss and Dromanns introduced the Kohonen's algorithm to realize this kind of topology-conserving mapping.

Model by Bhanot *et al.*

In 2005, a model was developed by Bhanot *et al.* to minimize only inter-task communication. They neglected the actual computing cost when placing tasks on processors linked by mesh or torus by the following model:

$$\min F = \sum_{i,j} C(i, j) H(i, j), \tag{5.5}$$

where $C(i,j)$ is communication data from domain i to j and $H(i,j)$ represents the smallest number of hops on BlueGene/L torus between processors allocated domains i and j. A simulated annealing technique is used to map n tasks to n processors on the supercomputer for minimizing the communication time.

5.3.4. *Mapping models and algorithms*

In 2008, Chen and Deng considered a more realistic and general case to model the mapping problem. Assume the application has already been appropriately decomposed as n subtasks and the computing load of each subtask is equal. The inter-subtask communication requirements thus can be described by the demand matrix $D_{n \times n}$ whose entry $D(t, t')$ is the required data from subtask t to subtask t'.

Under the condition that the communication of the parallel computer is heterogeneous, we can further assume that the heterogeneous properties can be described by a load matrix $L_{n \times m}$ whose entry $L(t, p)$ is the computation cost of the subtask t when t is executed on the processor p, and a supply matrix $S_{m \times m}$ whose entry $S(p, p')$ is the cost for communication between processors p and p', where n is the number of subtasks and m is the number of processors. With the assumptions above, we can formulate the mapping problem as a quadratic assignment problem in the following way.

Let $\{t_1, \ldots, t_n\}$ be the set of subtasks of the problem and $\{p_1, \ldots, p_m\}$ be the set of heterogeneous processors of the parallel computers to which the subtasks are assigned. In general, $n > m$. Let X_{tp} be the decision Boolean variable that is defined as,

$$x_{tp} = \begin{cases} 1, & \text{if subtask } t \text{ is assigned to processor } p \\ 0, & \text{otherwise.} \end{cases} \tag{5.6}$$

Let $Y_{tt'pp'}$ be the decision Boolean variable that is defined as,

$$y_{tt'pp'} = \begin{cases} 1, & \text{if subtasks } t, t' \text{ assigned to } p, p' \text{ respectively} \\ 0, & \text{otherwise.} \end{cases} \tag{5.7}$$

The total execution time can be expressed as,

$$\min \left\{ \sum_{t=1}^{n} \sum_{p=1}^{m} L(t, p) \cdot x_{tp} + \sum_{i=1}^{k} (D_i S)_{\max} \right\}. \tag{5.8}$$

Subject to,

$$\begin{cases} \displaystyle\sum_{p=1}^{m} x_{tp} = 1, & t = 1, \ldots, n \\[2mm] \displaystyle\sum_{t=1}^{n} x_{tp} \geq 1, & p = 1, \ldots, m \\[2mm] \displaystyle\sum_{t=1}^{n} x_{tp} \leq \left[\dfrac{A_p}{A_t} \times n \right], & p = 1, \ldots, m, \end{cases} \qquad (5.9)$$

where $k \leq n^2$ is the total number of the communication batch. The term $(D_i S)_{\max}$ represents the maximum value of the ith batch communication.

It can be seen that the mapping problem can be concluded as a minimization problem. When considering the scale of modern parallel computers, optimization techniques often fail for problems of this size. Several heuristic techniques have been developed for search in large solution spaces, such as simulated annealing (SA), genetic algorithm (GA), evolution strategies (ES), genetic simulated annealing (GSA) and Tabu search (TS).

Figure 5.15 illustrated some of the most important techniques for the mapping problem. Details of the each algorithm can be easily found in literatures.

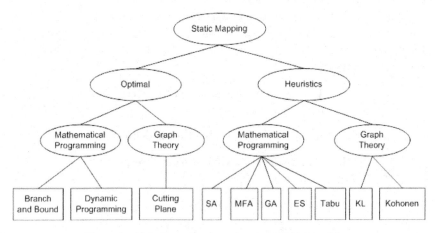

Fig. 5.15. Solution techniques for the mapping problem.

CHAPTER 6

LINEAR ALGEBRA

Linear algebra is one of the fundamental subjects of mathematics and the building block of various scientific and engineering applications such as fluid dynamics, image and signal processing, structural biology and many others. Because of its importance, numerical linear algebra has been paid massive amount of attention, for both serial and parallel computing systems, before and after the 1990s, respectively. After years of development, parallel linear algebra is among the most mature fields of the parallel computing applications. LINPACK and ScaLAPACK are their representative products and research publications by Dongarra, Golub, Greenbaum, Ortega and many others are a small subset of the rich publications in numerical linear algebra.

6.1. Problem Decomposition

A naïve way of working on parallel linear algebra is to copy all the data to be computed across the processors and perform the parallelization only in computing. In this situation, most of the parallelization is straightforward, since all information is simultaneously available to all processors.

However, the size of matrices in practical applications is so large that it not only requires intensive computing power, but also exceeds the memory capacity that one usually can get for a single node. For example, a moderate mesh of $1000 \times 1000 = 10^6$ points can induce a matrix of size $10^6 \times 10^6$ which means 10^{12} double precision data that roughly requires eight terabytes of memory. Hence, in this situation, decomposition across data is also required.

Consider a data matrix:

$$\begin{pmatrix} M_{11} & M_{12} & \cdots & M_{1n} \\ M_{21} & M_{22} & \cdots & M_{2n} \\ \vdots & \vdots & \ddots & \vdots \\ M_{m1} & M_{m2} & \cdots & M_{mn} \end{pmatrix} \tag{6.1}$$

where $mn = N$ and P is the number of processors and usually, $N > P$. Otherwise, if $N < P$, parallel computing would make little sense. In most analysis, we assume N/P to be an integer; however, it is not a big problem if it is not. In case where N/P is not an integer, we can patch "0" elements to make it an integer. For the array of processors, we can also arrange them in a processor matrix format:

$$\begin{pmatrix} P_{11} & P_{12} & \cdots & P_{1q} \\ P_{21} & P_{22} & \cdots & P_{2q} \\ \vdots & \vdots & \ddots & \vdots \\ P_{p1} & P_{p2} & \cdots & P_{pq} \end{pmatrix} \tag{6.2}$$

where $pq = P$.

Generally, there are four methods of decomposition that are common in practical use:

(1) Row partition.
(2) Column partition.
(3) Block partition.
(4) Scatter partition.

A fifth method of random decomposition, albeit rarely, is another possibility.

The following task assignment notation (TAN) of matrix decomposition is introduced by one of the authors in an earlier publication. The potential of such notation has not been fully exploited. In this notation, $m_{\alpha\beta}@P$ means that sub-matrix $m_{\alpha\beta}$ (which could be a single scalar number) is assigned to processor P. When carrying computations, communication logistics is visually evident. For example, when multiplying two square matrices in *four* processors,

$$\begin{aligned} C &= \begin{pmatrix} a_{11}@1 & a_{12}@2 \\ a_{21}@3 & a_{22}@4 \end{pmatrix} \begin{pmatrix} b_{11}@1 & b_{12}@2 \\ b_{21}@3 & b_{22}@4 \end{pmatrix} \\ &= \begin{pmatrix} c_{11} & c_{12} \\ c_{21} & c_{22} \end{pmatrix} \end{aligned} \tag{6.3}$$

where

$$
\begin{aligned}
c_{11} &= (a_{11}@1) \times (b_{11}@1) + (a_{12}@2) \times (b_{21}@3) \\
c_{12} &= (a_{11}@1) \times (b_{12}@2) + (a_{12}@2) \times (b_{22}@4) \\
c_{21} &= (a_{21}@3) \times (b_{11}@1) + (a_{22}@4) \times (b_{21}@3) \\
c_{22} &= (a_{21}@3) \times (b_{12}@2) + (a_{22}@4) \times (b_{22}@4)
\end{aligned}
\tag{6.4}
$$

Obviously, calculation of $(a_{11} \times b_{11})@1$ can be carried out on $P = 1$ without any communication. Calculation of $(a_{12}@2) \times (b_{21}@3)$ requires moving b_{21} from $P = 3$ to $P = 2$ or a_{12} from $P = 2$ to $P = 3$ before computing. The advantage is that all such communication patterns that could be excruciatingly complicated can now be mathematically manipulated. It is not difficult to quickly expect to exploit the newly introduced task assignment notation (TAN) for helping manipulate many other algorithms than the matrix multiplications. In fact, this TAN for matrix multiplication can guide us to express the algorithm by a PAT graph conveniently.

Row Partition: Assigning a subset of entire row to a processor.

$$
\begin{pmatrix}
M_{11}@1 & M_{12}@1 & \cdots & M_{1q}@1 \\
M_{21}@2 & M_{22}@2 & \cdots & M_{2q}@2 \\
\vdots & \vdots & \ddots & \vdots \\
M_{p1}@p & M_{p2}@p & \cdots & M_{pq}@p
\end{pmatrix}
\tag{6.5}
$$

Column Partition: Assigning a subset of entire column to a processor.

$$
\begin{pmatrix}
M_{11}@1 & M_{12}@2 & \cdots & M_{1q}@p \\
M_{21}@1 & M_{22}@2 & \cdots & M_{2q}@p \\
\vdots & \vdots & \ddots & \vdots \\
M_{p1}@1 & M_{p2}@2 & \cdots & M_{pq}@p
\end{pmatrix}
\tag{6.6}
$$

Block Partition: Assigning a continuous block of sub-matrix to a process, a typical 2D matrix partition.

$$
\begin{pmatrix}
M_{11}@1 & M_{12}@1 & \cdots & M_{1q}@3 \\
M_{21}@1 & M_{22}@1 & \cdots & M_{2q}@3 \\
M_{31}@2 & M_{32}@2 & \cdots & M_{3q}@k \\
M_{41}@2 & M_{42}@2 & \cdots & M_{4q}@k \\
\vdots & \vdots & \ddots & \vdots \\
M_{p1}@4 & M_{p2}@4 & \cdots & M_{pq}@n
\end{pmatrix}
\tag{6.7}
$$

Scattered Partition: Assigning a subset of scattered sub-matrices forming a particular pattern to a processor, a type of 2D partition.

$$\begin{pmatrix} M_{11}@1 & M_{12}@2 & \cdots & M_{1q}@x \\ M_{21}@3 & M_{22}@4 & \cdots & M_{2q}@y \\ M_{31}@1 & M_{32}@2 & \cdots & M_{3q}@x \\ M_{41}@3 & M_{42}@4 & \cdots & M_{4q}@y \\ \vdots & \vdots & \ddots & \vdots \\ M_{p1}@1 & M_{p2}@2 & \cdots & M_{pq}@n \end{pmatrix} \qquad (6.8)$$

The properties of these decompositions vary widely. The table below lists their key features:

Method	Properties
Row partition and column partition	• 1D decomposition. • Easy to implement. • Relatively high communication costs.
Block partition and scattered partition	• 2D decomposition. • More complication to implement. • Lower communication costs. Block partition: Mismatch with underlying problem mesh. Scattered partition: Potential match with underlying problem mesh.

6.2. Matrix Operations

Matrix-matrix multiplication is the fundamental operation in linear algebra and matrix-vector multiplication is the basic operation in matrix-matrix multiplication. To learn matrix-matrix multiplication, we start from matrix-vector multiplication.

6.2.1. *Matrix-vector multiplications*

When multiplying a matrix and a vector, we compute:

$$Ab = c \qquad (6.9)$$

Method I: Replicating the vector on all processors

Matrix A: Row partition matrix A according to Eq. (6.5):

$$\begin{pmatrix} A_{11}@1 & A_{12}@1 & \cdots & A_{1q}@1 \\ A_{21}@2 & A_{22}@2 & \cdots & A_{2q}@2 \\ \vdots & \vdots & \ddots & \vdots \\ A_{p1}@p & A_{p2}@p & \cdots & A_{pq}@p \end{pmatrix}$$

Vector b: The elements of vector b are given to each processor i for each integer $i \in [1, p]$:

$$\begin{pmatrix} b_1@i \\ b_2@i \\ b_3@i \\ \vdots \\ b_x@i \end{pmatrix} \tag{6.10}$$

We can obtain the results on individual processors for the resulting vector c. The PAT graph for matrix-vector multiplication Method I is as given in Fig. 6.1.

Timing Analysis: Each processor now works on a $n \times \frac{n}{p}$ matrix with a vector of length n. The time on each processor is given by the following. On P processors,

$$T_{\text{comp}}(n, p) = cn \times \frac{n}{p} \tag{6.11}$$

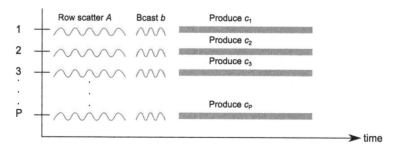

Fig. 6.1.   PAT graph for matrix-vector multiplication Method I.

The communication time to form the final vector is:

$$T_{\text{comm}}(n, p) = dp \times \frac{n}{p} \qquad (6.12)$$

Thus, the speed up is:

$$S(n, p) = \frac{T(n, 1)}{T_{\text{comp}}(n, p) + T_{\text{comm}}(n, p)} = \frac{P}{1 + \frac{cp}{n}} \qquad (6.13)$$

Remarks

(1) Overhead is proportional to p/n and speedup increases with n/p.
(2) This method is quite easy to implement.
(3) Memory is not parallelized (redundant use of memory).
(4) Applications where such decompositions are useful are rare.

Method II: Row partition both A and B

Matrix A: Row partition matrix A according to Eq. (6.5):

$$\begin{pmatrix} A_{11}@1 & A_{12}@1 & \cdots & A_{1q}@1 \\ A_{21}@2 & A_{22}@2 & \cdots & A_{2q}@2 \\ \vdots & \vdots & \ddots & \vdots \\ A_{p1}@p & A_{p2}@p & \cdots & A_{pq}@p \end{pmatrix}$$

Vector b: In this case, the element b_i is given to the ith processor:

$$\begin{pmatrix} b_1@1 \\ b_2@2 \\ b_3@3 \\ \vdots \\ b_x@p \end{pmatrix} \qquad (6.14)$$

Step 1: Each processor multiplies its diagonal element of A with its element of b.

$$@1: \quad A_{11} \times b_1$$
$$@2: \quad A_{22} \times b_2$$
$$\vdots \qquad \vdots$$
$$@p: \quad A_{pp} \times b_p$$

Step 2: "Roll up" in the processor vector.

Vector b:

$$\begin{pmatrix} b_2@1 \\ b_3@2 \\ b_4@3 \\ \vdots \\ b_1@p \end{pmatrix} \tag{6.15}$$

Then, multiply the next off-diagonal elements with the local vector elements:

$$\begin{array}{ll} @1: & A_{12} \times b_2 \\ @2: & A_{23} \times b_3 \\ \vdots & \vdots \\ @p: & A_{p1} \times b_1 \end{array}$$

Step 3: Repeat Step 2 until all elements of b have visited all processors.

The PAT graph for matrix-vector multiplication method II is as in Fig. 6.2.

Timing Analysis: Each processor now works on a $n \times \frac{n}{p}$ matrix with a vector on length n. The time on each processor is given by the following. On P processors,

$$T_{\text{comp}}(n,p) = cn \times \frac{n}{p} \tag{6.16}$$

The communication time to roll up the elements of vector b and form the final vector is,

$$T_{\text{comm}}(n,p) = dp \times \frac{n}{p} \tag{6.17}$$

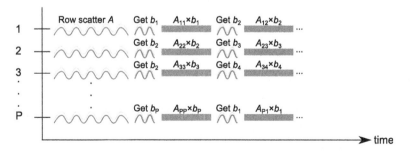

Fig. 6.2. PAT graph for matrix-vector multiplication Method II.

Thus, the speedup is,

$$S(n,p) = \frac{T(n,1)}{T_{\text{comp}}(n,p) + T_{\text{comm}}(n,p)} = \frac{P}{1 + \frac{c'p}{n}} \qquad (6.18)$$

Remarks

(1) Overhead is proportional to p/n and speedup increases with n/p.
(2) This method is not as easy to implement.
(3) Memory use, as well as communication, is parallelized.
(4) Communication cost is higher than method I.

Method III: Column partition A and row partition b

Matrix A: Column partition matrix A according to Eq. (6.6):

$$\begin{pmatrix} A_{11}@1 & A_{12}@2 & \cdots & A_{1q}@p \\ A_{21}@1 & A_{22}@2 & \cdots & A_{2q}@p \\ \vdots & \vdots & \ddots & \vdots \\ A_{p1}@1 & A_{p2}@2 & \cdots & A_{pq}@p \end{pmatrix}$$

Vector b: The element b_i is given to the ith processor as in Eq. (6.14).

Step 1: Each processor i, multiplies A_{i1} with b_i.

$$\begin{array}{ll} @1: & A_{11} \times b_1 \\ @2: & A_{12} \times b_2 \\ \vdots & \vdots \\ @p: & A_{pp} \times b_p \end{array}$$

The results are then "lumped" to form the element c_1 of the vector c.

The PAT graph for matrix-vector multiplication method III is as in Fig. 6.3.

Step 2: Repeat Step 1 for all rows to form the complete resulting vector c.

Timing Analysis: Each processor now works on a $n \times n/p$ matrix with a vector on length n. The time on each processor is given by the following.

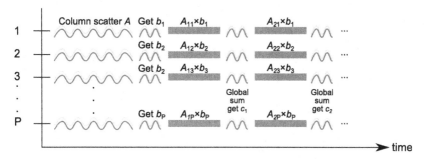

Fig. 6.3. PAT graph for matrix-vector multiplication Method III.

On P processors,

$$T_{\text{comp}}(n, p) = cn \times \frac{n}{p} \tag{6.19}$$

The communication required to lump all elements to form one element of c is given by:

$$T_{\text{comm}}(n, p) = d''p \times \frac{n}{p} \tag{6.20}$$

Thus, the speedup is,

$$S(n, p) = \frac{T(n, 1)}{T_{\text{comp}}(n, p) + T_{\text{comm}}(n, p)} = \frac{P}{1 + \frac{c'p}{n}} \tag{6.21}$$

Remarks

(1) Overhead is proportional to p/n and speedup increases with n/p.
(2) This method is not as easy to implement.
(3) Memory use, as well as communication, is parallelized.
(4) Communication cost is higher than method I.

6.2.2. *Matrix-matrix multiplications*

When multiplying two matrices, we compute

$$AB = C \tag{6.22}$$

where,

$$A = \begin{pmatrix} A_{11} & A_{12} & \cdots & A_{1n} \\ A_{21} & A_{22} & \cdots & A_{2n} \\ \vdots & \vdots & \ddots & \vdots \\ A_{m1} & A_{m2} & \cdots & A_{mn} \end{pmatrix} \tag{6.23}$$

$$B = \begin{pmatrix} B_{11} & B_{12} & \cdots & B_{1k} \\ B_{21} & B_{22} & \cdots & B_{2k} \\ \vdots & \vdots & \ddots & \vdots \\ B_{n1} & B_{n2} & \cdots & B_{nk} \end{pmatrix} \tag{6.24}$$

The complexity in multiplying A and B is $O(n^3)$. Implementation of matrix multiplication algorithms on a serial computer requires only four meaningful lines of program as shown in Fig. 6.4:

Implementation of matrix multiplication algorithms on a parallel computer, however, is an entirely different matter.

Method I: Row-Column partition (Ring Method) (Fig. 6.5)

```
for ( I = 1; I <= n; I++) {
    for ( J = 1; J <= n; J++) {
        for ( K = 1; K <= n; K++) {
            C[I][J] += A[I][K] * B[K][j];
        }
    }
}
```

Fig. 6.4.   This is the sequential code for matrix multiplication.

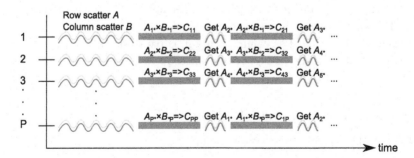

Fig. 6.5.   PAT graph for matrix-matrix multiplication Method I.

Step 0: Row partition matrix A and column partition matrix B.

$$A = \begin{pmatrix} A_{11}@1 & A_{12}@1 & \cdots & A_{1n}@1 \\ A_{21}@2 & A_{22}@2 & \cdots & A_{2n}@2 \\ A_{31}@3 & A_{32}@3 & \cdots & A_{3n}@3 \\ \vdots & \vdots & \ddots & \vdots \\ A_{m1}@m & A_{m2}@m & \cdots & A_{mx}@m \end{pmatrix}$$

$$B = \begin{pmatrix} B_{11}@1 & B_{12}@2 & \cdots & B_{1k}@k \\ B_{21}@1 & B_{22}@2 & \cdots & B_{2k}@k \\ B_{31}@1 & B_{32}@2 & \cdots & B_{3k}@k \\ \vdots & \vdots & \ddots & \vdots \\ B_{n1}@1 & B_{n2}@2 & \cdots & B_{nk}@k \end{pmatrix}$$

Step 1: Multiply row 1 of A with column 1 of B to create c_{11} by processor 1, multiply row 2 of A with column 2 of B to create c_{22} by processor 2, and so on. Therefore, the diagonal elements of the matrix c are created in parallel.

$$C = \begin{pmatrix} C_{11} & & & \\ & C_{22} & & \\ & & \ddots & \\ & & & C_{mm} \end{pmatrix}$$

where,

$$@1: \quad C_{11} = \sum_{i=1}^{n} A_{1i} B_{i1}$$

$$@2: \quad C_{22} = \sum_{i=1}^{n} A_{2i} B_{i2}$$

$$\vdots \qquad \vdots$$

$$@m: \quad C_{mm} = \sum_{i=1}^{n} A_{1i} B_{im}$$

Step 2: Roll up all rows in A by one processor unit to form A then multiply A's corresponding rows with B's and columns to form the first off diagonal

C elements. Now, A looks like this:

$$A = \begin{pmatrix} A_{21}@1 & A_{22}@1 & \cdots & A_{2n}@1 \\ A_{31}@2 & A_{32}@2 & \cdots & A_{3n}@2 \\ A_{41}@3 & A_{42}@3 & \cdots & A_{4n}@3 \\ \vdots & \vdots & \ddots & \vdots \\ A_{11}@m & A_{12}@m & \cdots & A_{1n}@m \end{pmatrix}$$

and C looks like,

$$C = \begin{pmatrix} C_{11} & & & C_{1m} \\ C_{21} & C_{22} & & \\ & \ddots & \ddots & \\ & & C_{m,m-1} & C_{mm} \end{pmatrix}$$

where

$$@1: \quad C_{21} = \sum_{i=1}^{n} A_{2i}B_{i1}$$

$$@2: \quad C_{32} = \sum_{i=1}^{n} A_{3i}B_{i2}$$

$$\vdots \qquad \vdots$$

$$@m: \quad C_{1m} = \sum_{i=1}^{n} A_{1i}B_{im}$$

Step 3: Repeat Step 2 until all rows of A have passed through all processors.

Timing Analysis: On one processor,

$$T_{\text{comp}}(n, 1) = cn^3 t_{\text{comp}} \tag{6.25}$$

where t_{comp} is the time needed to perform unit computations, such as multiplying a pair of numbers.

On P processors, the total cost is:

$$T(n, p) - T_{\text{comp}}(n, p) + T_{\text{comm}}(n, p) \tag{6.26}$$

The communication cost to roll the rows of A is,

$$T_{\text{comm}}(n, p) = (p - 1) \left(\frac{n^2}{p} \right) t_{\text{comm}} \approx n^2 t_{\text{comm}} \qquad (6.27)$$

The computation cost for multiplying a matrix of size n/p with a matrix of size n is:

$$T_{\text{comp}}(n, p) = p^2 c \left(\frac{n}{p} \right)^3 t_{\text{comp}} = \frac{cn^3 t_{\text{comp}}}{p} \qquad (6.28)$$

Thus, the speedup is:

$$S(n, p) = \frac{T(n, 1)}{T_{\text{comp}}(n, p) + T_{\text{comm}}(n, p)} = \frac{P}{1 + \frac{c'p}{n}} \qquad (6.29)$$

Remarks

(1) Overhead is proportional to p/n and speedup increases with n/p.
(2) Overhead is proportional to $T_{\text{comm}}/T_{\text{comp}}$ and speedup decreases while $T_{\text{comm}}/T_{\text{comp}}$ increases.
(3) The previous two comments are universal for parallel computing.
(4) Memory is parallelized.

Method II: Broadcast, Multiply, Roll (BMR Method)

Step 0: Block partition both A and B.

$$A = \begin{pmatrix} A_{11}@1 & A_{12}@2 & A_{13}@3 \\ A_{21}@4 & A_{22}@5 & A_{23}@6 \\ A_{31}@7 & A_{32}@8 & A_{33}@9 \end{pmatrix}$$

$$B = \begin{pmatrix} B_{11}@1 & B_{12}@2 & B_{13}@3 \\ B_{21}@4 & B_{22}@5 & B_{23}@6 \\ B_{31}@7 & B_{32}@8 & B_{33}@9 \end{pmatrix}$$

Step 1: Broadcast all diagonal elements of A to the individual processors on each row. In other words, the elements of A will be distributed to the processors in the following way:

$$\begin{pmatrix} A_{11}@1 & A_{11}@2 & A_{11}@3 \\ A_{22}@4 & A_{22}@5 & A_{22}@6 \\ A_{33}@7 & A_{32}@8 & A_{33}@9 \end{pmatrix}$$

Step 2: Multiply row 1 of A with row 1 of B, multiply row 2 of A with row 2 of B, etc., to produce part of matrix C.

@1: $A_{11} \times B_{11} \Rightarrow C_{11}(\text{partial})$

@2: $A_{11} \times B_{12} \Rightarrow C_{12}(\text{partial})$

@3: $A_{11} \times B_{13} \Rightarrow C_{13}(\text{partial})$

@4: $A_{22} \times B_{21} \Rightarrow C_{21}(\text{partial})$

@5: $A_{22} \times B_{22} \Rightarrow C_{22}(\text{partial})$

@6: $A_{22} \times B_{23} \Rightarrow C_{23}(\text{partial})$

@7: $A_{33} \times B_{31} \Rightarrow C_{31}(\text{partial})$

@8: $A_{33} \times B_{32} \Rightarrow C_{32}(\text{partial})$

@9: $A_{33} \times B_{33} \Rightarrow C_{33}(\text{partial})$

Step 3: Broadcast the next diagonal elements of A, roll up the rows of B and then multiply as in Step 2. Matrix A and B become

$$A = \begin{pmatrix} A_{12}@1 & A_{12}@2 & A_{12}@3 \\ A_{23}@4 & A_{23}@5 & A_{23}@6 \\ A_{31}@7 & A_{31}@8 & A_{31}@9 \end{pmatrix}$$

$$B = \begin{pmatrix} B_{21}@1 & B_{22}@2 & B_{23}@3 \\ B_{31}@4 & B_{32}@5 & B_{33}@6 \\ B_{11}@7 & B_{12}@8 & B_{13}@9 \end{pmatrix}$$

@1: $A_{12} \times B_{21} \Rightarrow C_{11}(\text{partial})$

@2: $A_{12} \times B_{22} \Rightarrow C_{12}(\text{partial})$

@3: $A_{12} \times B_{23} \Rightarrow C_{13}(\text{partial})$

@4: $A_{23} \times B_{31} \Rightarrow C_{21}(\text{partial})$

@5: $A_{23} \times B_{32} \Rightarrow C_{22}(\text{partial})$

@6: $A_{23} \times B_{33} \Rightarrow C_{23}(\text{partial})$

@7: $A_{31} \times B_{11} \Rightarrow C_{31}(\text{partial})$

@8: $A_{31} \times B_{12} \Rightarrow C_{32}(\text{partial})$

@9: $A_{31} \times B_{13} \Rightarrow C_{33}(\text{partial})$

Step 4: Repeat Step 3 until all B elements have traveled through all processors, and then gather all the elements of C (Fig. 6.6).

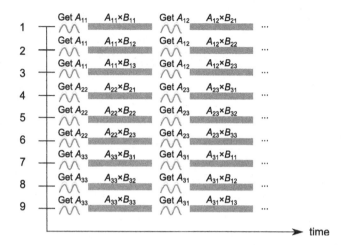

Fig. 6.6.   PAT graph for matrix-matrix multiplication Method II.

Timing Analysis: On one processor,

$$T_{\text{comp}}(n, 1) = 2n^3 t_{\text{comp}} \tag{6.30}$$

where t_{comp} is the time needed to perform unit computations, such as multiplying a pair of numbers.

On P processors, the total cost, including the broadcast (communication), sub-matrix multiply, and rolling up of the matrix elements, is:

$$T(n, p) - T_B(n, p) + T_M(n, p) + T_R(n, p) \tag{6.31}$$

The broadcast cost is:

$$T_B(n, p) = m^2 t_{\text{comm}} + (q - 2) t_{\text{startup}} \tag{6.32}$$

where $m = n/p$.

The cost of sub-matrix multiplication is:

$$T_M(n, p) = 2m^3 t_{\text{comp}} \tag{6.33}$$

The cost of rolling up the rows of the matrix is:

$$T_R(n, p) = 2m^2 t_{\text{comm}} \tag{6.34}$$

Thus, the speedup is:

$$S(n, p) = \frac{T(n, 1)}{T_B(n, p) + T_M(n, p) + T_R(n, p)} \tag{6.35}$$

and the overhead is:

$$h(n, p) = \left(\frac{1}{np} + \frac{P - 2q}{2\pi^2}\right)\left(\frac{t_{\text{comm}}}{t_{\text{comp}}}\right) \tag{6.36}$$

Remarks

(1) Overhead is proportional to p/n and speedup increases with n/p.
(2) Overhead is proportional to $T_{\text{comm}}/T_{\text{comp}}$ and speedup increases while $T_{\text{comm}}/T_{\text{comp}}$ decreases.
(3) The above two comments apply commonly to many parallel algorithms.
(4) Memory is parallelized.

Method III: Strassen Method

Matrix multiplication requires $O(N^3)$ "×" operations. For example, suppose we perform $AB = C$ where

$$A = \begin{pmatrix} a_{11} & a_{12} \\ a_{21} & a_{22} \end{pmatrix}, \quad B \begin{pmatrix} b_{11} & b_{12} \\ b_{21} & b_{22} \end{pmatrix}$$

How many multiplication operations are required? The answer is $2^3 = 8$.

$$\begin{aligned}
a_{11} \times b_{11} + a_{12} \times b_{21} &\to c_{11} \\
a_{12} \times b_{12} + a_{12} \times b_{22} &\to c_{12} \\
a_{21} \times b_{11} + a_{22} \times b_{21} &\to c_{21} \\
a_{21} \times b_{12} + a_{22} \times b_{22} &\to c_{22}
\end{aligned} \tag{6.37}$$

The same applies to a 3×3 matrix. There are $3^3 = 27$ operations.

Now, if we were to rearrange the operations in the following way,

$$\begin{aligned}
S_1 &= (a_{11} + a_{22}) \times (b_{11} + b_{22}) \\
S_2 &= (a_{21} + a_{22}) \times b_{11} \\
S_3 &= a_{11} \times (b_{12} - b_{22}) \\
S_4 &= a_{22} \times (-b_{11} + b_{21}) \\
S_5 &= (a_{11} + a_{12}) \times b_{22} \\
S_6 &= (-a_{11} + a_{21}) \times (b_{11} + b_{12}) \\
S_7 &= (a_{12} - a_{22}) \times (b_{21} + b_{22})
\end{aligned} \tag{6.38}$$

we can express c_{ij} in terms of S_k,

$$
\begin{aligned}
C_{11} &= S_1 + S_4 - S_5 + S_7 \\
C_{12} &= S_3 + S_5 \\
C_{21} &= S_2 + S_4 \\
C_{22} &= S_1 + S_3 - S_2 + S_5
\end{aligned}
\tag{6.39}
$$

This reduces the number of multiplications from eight to seven. For large matrices, we can always reduce it to 7/8 and then get the "×" computing time from $N \log_2 8 = 3N$ to $\log_2 7 \approx 2.8N$. Of course, we have many more additions, namely five times as many. For large N, however, the cost of the "×" operations will outweigh the cost of the "+" operations.

6.3. Solution of Linear Systems

A system of linear equations, or a linear system, can be written in the matrix form

$$
A\mathbf{x} = b \tag{6.40}
$$

where $\mathbf{x}$ is the vector of unknown variables and A is $n \times n$ coefficient matrix where $n^2 = N$. Generally, there are two ways of solving the system: direct methods and iterative methods. Direct methods offer exact solution for the system while iterative methods give fast approximation especially for sparse matrices.

6.3.1. *Direct methods*

For dense linear systems, it is rare to find better methods than the so-called direct methods, although it will cost as much as $O(N^3)$ of computation. Here we discuss the classical method of Gaussian elimination. In this method, we transform the coefficient matrix A in (6.40) to an upper triangular matrix U and then solve the system by back-substitution.

Simple Gaussian Elimination

The sequential algorithm for Gaussian elimination is actually straightforward. Figure 6.7 gives the pseudo-code that constitutes three nested loops for Gaussian elimination.

```
u = a
for i = 1 to n - 1
    for j = i+ 1 to n
        l = u(j,i)/u(i,i)
        for k = i to n
            u(j,k) = u(j,k) - l * u(i,k)
        end
    end
end
```

Fig. 6.7. Pseudo-code for sequential Gaussian elimination.

From the pseudo-code, it is easy to conclude that the operation count for the Gaussian elimination is:

$$T_{\text{comp}}(n, 1) = \frac{2}{3}n^3 t_{\text{comp}} \qquad (6.41)$$

where t_{comp} is the time for unit computation.

Back Substitution

After the elimination, the system became of the form

$$Ux = b \qquad (6.42)$$

where U is an upper-triangular matrix. To finally solve the system, we have to perform the back substitution. That is, the last element of the vector of the unknown variables can be directly solved from the last equation, and by substituting it into the rest of the equations, it can generate a similar system of one lower degree. By repeating the process, x can thus be solved.

The method of parallelization for solving the linear system depends on the way the matrix is decomposed. In practice, it is unnatural to consider scattered partition for solving linear systems, so we only discuss the three ways of matrix decompositions.

Row Partition

Row partition is among the most straightforward methods that parallelizes the inner most loop. For simplicity, when demonstrating this method, we assume the number of processors to be equal to the number of dimensions i.e. $p = n$, but it is quite easy to port it into situations that $p < n$.

Step 0: Partition the matrix according to its rows.
Step 1: Broadcast the first row to other rows.

Step 2: For every element in row i subtract the value of the corresponding element in first row multiplied by $\frac{A(0,0)}{A(i,0)}$. By doing this, the elements in the first column of the matrix become all zero.

Step 3: Repeat Steps 1 and 2 on the right-lower sub matrix until the system becomes an upper triangle matrix.

(0,0)	(0,1)	(0,2)	(0,3)	(0,4)
0	(1,1)	(1,2)	(1,3)	(1,4)
0	(2,1)	(2,2)	(2,3)	(2,4)
0	(3,1)	(3,2)	(3,3)	(3,4)
0	(4,1)	(4,2)	(4,3)	(4,4)

In this method, as we can see, every step has one less node involving in the computation. This means the parallelization is largely imbalanced. The total computation time can be written as:

$$T_{\text{comp}}(n,p) = \frac{n^3}{p} t_{\text{comp}} \tag{6.43}$$

where t_{comp} is the time for unit operation. Compare with the serial case (6.42), we know that roughly 1/3 of the computing power is wasted in waiting.

During the process, the right-hand side can be treated as the last column of the matrix. Thus, it can be distributed among the processor and thus no special consideration is needed.

For the back substitution, we can parallelize by the following procedure:

Step 1: Solve the last unsolved variable in the last processor.

Step 2: Broadcast the value of the solved variable.

Step 3: Every processor plugs this variable and updates its corresponding row.

Gaussian Elimination with Partial Pivoting

Like the serial cases, the algorithm might introduce big s when subtracting two very close numbers. Partial pivoting is necessary in the practical use.

That is, at each step, we find the row of the sub-matrix that has the largest first column elements and calculate based on this row instead of the first one. This requires finding the largest elements among processors that adds more communication in the algorithm. An alternative way to solve this problem is to use the column partition.

Column Partition

Column partition gives an alternative way of doing 1D partition. In this method, less data will be transferred in each communication circle. This gives smaller communication overhead and easier for pivoting. However, the problem with load imbalance still existed in this method.

Step 0: Partition the matrix according to its columns.
Step 1: Determine the largest element in the first column (pivot).
Step 2: Broadcast the pivot and its row number to other processors.
Step 3: Every processor updates its other elements.
Step 4: Repeat Steps 1 to 3 until finish.

This method has same overall computation time but much less communication when pivoting is needed.

Block Partition

We illustrate the idea by a special example 20×20 matrix solved by 16 processors. The idea can be easily generalized to more processors with a large matrix.

We perform the elimination with the following three steps.

Step 1: Broadcast column 0 to other "column" processors. Note, at this step, only $\sqrt{p-1}$ processors need to be sent the column 0 elements.
Step 2: Normalize column 0, then broadcast row 0 to the other "row" processors. Note, at this step, only $\sqrt{p-1}$ processors need to be sent to the row 0 elements. Eliminate column 0.
Step 3: Move to the next row and column and repeat Steps 1 and 2 until the last matrix element is reached.

Remarks

(1) There are many other schemes for solving $Ax = b$; for example, row partition or column partition.
(2) This method is, in fact, quite efficient but the load imbalance problem persists.

(3) It is quite unnatural to scatter matrix elements to processors in such a "strange" fashion as demonstrated.

(4) This method is also difficult to implement.

LU Factorization

In the practical applications, people usually need to solve a linear system

$$Ax = b$$

for multiple times with different b while A is constant. In this case, a pre-decomposed matrix that offers fast calculation of solution with different b's is desirable. To achieve this, LU decomposition is a good method.

Since LU factorization is actually closely related with Gaussian elimination, the algorithm is almost identical. The L matrix can be obtained by saving the l value in Fig. 6.7 and the resulting matrix in Gaussian elimination is just the U matrix. Thus, the pseudo code (Fig. 6.8) for LU factorization looks like the following.

Thus, it is easy to see that the parallelization of the LU factorization is very similar to that of the Gaussian elimination as shown in Fig. 6.9.

6.3.2. *Iterative methods*

There are rich families of serial algorithms for sparse linear systems, but designing efficient parallel algorithms for such systems has been a challenge, due to the lack of parallelism in these systems.

General Description

The P processors are used to solve sparse tri- (3-), 5-, and 7-diagonal systems (also known as banded systems) of the form $Ax = b$. For example,

```
u = a
for i = 1 to n - 1
    for j = i+ 1 to n
        l(j,i) = u(j,i)/u(i,i)
        for k = i to n
            u(j,k) = u(j,k) - l(j,i) * u(i,k)
        end
    end
end
```

Fig. 6.8.   Pseudo-code for sequential Gaussian elimination.

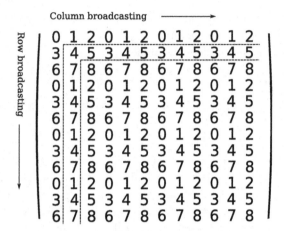

Fig. 6.9. This is an example of parallel gaussian elimination, i.e. a parallel algorithm used to solve $Ax = b$. The numbers shown in the matrix are the processor IDs. They indicate which matrix element is stored in which processor. This example shows a matrix of size 20×20 to be solved by 16 processors. Obviously, this choice can be easily generalized.

we may have

$$A = \begin{pmatrix} a_{11} & a_{12} & & \\ a_{21} & a_{22} & a_{23} & \\ & a_{32} & a_{33} & a_{34} \\ & & a_{43} & a_{44} \end{pmatrix}, \quad x = \begin{pmatrix} x_1 \\ x_2 \\ x_3 \\ x_4 \end{pmatrix}, \quad b = \begin{pmatrix} b_1 \\ b_2 \\ b_3 \\ b_4 \end{pmatrix} \tag{6.44}$$

Classical Techniques

- Relaxation
- Conjugate-gradient
- Minimal-residual

$$u^{(k)} = T u^{(k-1)} + C \tag{6.45}$$

where $u^{(0)}$ are initial values and T is the iteration matrix, tailored by the structure of A.

6.3.3. *ADI*

The alternating direction implicit (ADI) method was first used in 1950s by Peaceman and Rachford for solving parabolic PDEs. Since then, it has been widely used in many other applications as well.

Varga has refined and extended the discussions of this method. Its implementations on different parallel computers are discussed in various books and papers on parallel linear algebra.

We have the following implementations:

I. ADI on a 1D array of processors.

II. ADI on a 2D mesh of processors.

III. ADI on other architectures.

CHAPTER 7

DIFFERENTIAL EQUATIONS

Another family of numerical algorithms involves solution of differential equations. Everything starts from the basics of numerical integration and differentiation. This chapter will discuss solutions of several classic linear partial differential equations after introducing the integration methods.

7.1. Integration and Differentiation

As one of the standard embarrassingly parallel problems, numerical integration is CPU time-consuming, particularly in high dimensions. Three integration methods for one-dimensional integrands are shown here and they can be easily generalized to higher dimensions.

7.1.1. *Riemann summation for integration*

Given the integral

$$I = \int_a^b f(x)dx \qquad (7.1)$$

slice the interval [a, b] into N mesh blocks, each of equal size $\Delta x = \frac{b-a}{N}$, shown in (7.1). Suppose $x_i = a + (i - 1)\Delta x$, the integral is approximated as in (Fig. 7.1):

$$I \approx \sum_{i=1}^{N} f(x_i)\Delta x \qquad (7.2)$$

This is the Riemann summation approximation of the integral. More sophisticated approximation techniques, e.g. the Trapezoid Rule, Simpson's Rule, etc. can also be easily implemented. The PAT graph can be drawn as shown in Fig. 7.2.

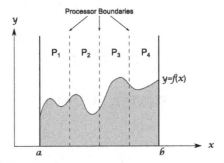

Fig. 7.1. $I = \int_a^b f(x)dx$. In the Riemann sum, the totall area A under the curve $f(x)$ is the approximation of the integral I. With 2 processors j and $j + 1$, we can compute $A = A_j + A_{j+1}$ completely in parallel. Each processor only needs to know its starting and ending points, in addition to the integrand, for partial integration. At the end, all participating processors perform a global summation to add up partial integrals obtained by individual processors.

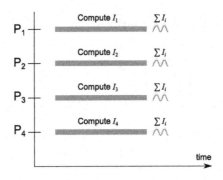

Fig. 7.2.   PAT graph for integration.

7.1.2. *Monte Carlo method for integration*

Unlike the deterministic Riemann summation, the computational load of the Monte Carlo integration does not grow when the number of dimensions grows. In fact, the convergence rate of Monte Carlo integration only depends on the number of trail evaluated. That is, for integration

$$I = \int_D g(x)dx \tag{7.3}$$

we can obtain an approximation as:

$$\hat{I}_m = \frac{1}{m}[g(x^{(1)}) + \cdots + g(x^{(m)})] \tag{7.4}$$

where $x^{(i)}$'s are random samples from D. According to the central limit theorem, the error term,

$$\sqrt{m}(\hat{I}_m - I) \to N(0, \sigma^2) \tag{7.5}$$

where $\sigma^2 = \text{var}\{g(x)\}$. This means the convergence rate of the Monte Carlo method is $O(m^{-1/2})$, regardless of the dimension of D. This property gives Monte Carlo method an advantage when integrating in high dimensions.

7.1.3. Simple parallelization

The most straightforward parallelization of Monte Carlo method is to let every processor generate its own random values $g(x_p^{(i)})$. Communication occurs once at the end of calculation for aggregating all the local approximations and produces the final result. Thus, the parallel efficiency will tend to close to 100%.

However, the true efficiency in terms of overall convergence is not achievable in practice. The more processors used in the calculation, the more points of evaluation is needed for the same level of accuracy. This is mainly caused by the collision of the random sampling (i.e. repetition of same number in different processors). due to the pseudo-random number generator commonly used in computing.

With $P = 1$ processor making m samples, the error converges as:

$$\varepsilon_1 \propto O\left(\frac{1}{\sqrt{m}}\right).$$

With P processors making m samples each, the error convergence is expected to be,

$$\varepsilon_P \propto O\left(\frac{1}{\sqrt{mP}}\right)$$

In fact, we are not precisely sure how the overall convergence varies with increasing number of processors. With a variation of the sampling, we may resolve the random number collision problem. For example, we slice the integration domain D into P sub-domains and assign each sub-domain to a processor. Each processor is restricted to sample in its designated sub-domain. The rest is similar to method I: Ordinary differential equations.

7.2. Partial Differential Equations

We discuss solutions of several hyperbolic, parabolic and elliptical differential equations, describe the algorithms, and analyze their performances.

7.2.1. *Hyperbolic equations*

Wave equations are representative of hyperbolic equations, which are the easiest differential equations to parallelize due to their inherent nature of locality. Thus, there exist many methods to solve these equations; finite-difference or finite-element methods are the major ones. Each is either an explicit scheme or implicit depending on the method chosen. The choice may depend on the requirements for the solution, or it may be dictated by the property of the partial differential equation (PDE).

When it is implicit, we will have an equation $AX = b$ to solve. Very often A is a spare matrix, thus an indirect method or iterative method is used. When A is dense, a direct method such as LU decomposition or Gaussian elimination is used. The simplest method is an explicit scheme, as those methods can be parallelized more easily.

1D Wave equation:

We first demonstrate parallel solution of 1D wave equations with the simplest method, i.e. an explicit finite difference scheme. Consider the following 1D wave equation:

$$\begin{cases} u_{tt} = c^2 u_{xx}, & \forall t > 0 \\ \text{Proper BC and IC} \end{cases} \tag{7.6}$$

with conveniently given initial and boundary conditions.

It is very easy to solve this equation on sequential computers, but on a parallel computer, it is a bit more complex. We perform finite difference operation (applying central differences on both sides):

$$\frac{u_i^{k+1} + u_i^{k-1} - 2u_i^k}{\Delta t^2} = \frac{u_{i+1}^k + u_{i-1}^k - 2u_i^k}{\Delta x^2} \tag{7.7}$$

where $u_{i\pm1}^{k\pm1} = u(x \pm \Delta x, t \pm \Delta t)$. Thus, we get an update scheme:

$$u_i^{k+1} = f(u_i^{k-1} u_i^k, \ldots, u_{i+1}^k u_{i-1}^k) \tag{7.8}$$

After decomposing the computational domain into sub-domains, each processor is given a sub-domain to solve the equation. When performing updates for the interior points, each processor can behave like a serial processor. However, when a point on the processor boundary needs updating, a point from a neighboring processor is needed. This requires communication. Most often, this is done by building buffer zones for the physical sub-domains. After updating its physical sub-domain, a processor will request that its neighbors (2 in 1D, 8 in 2D, and 26 in 3D) send their boarder mesh point solution values (the number of mesh point solution values sent depends on the numerical scheme used) and in the case of irregular and adaptive grid, the point coordinates. With this information, this processor then builds a so-called virtual sub-domain that contains the original physical sub-domain surrounded by the buffer zones communicated from other processors (Figs. 7.3 and 7.4).

Performance analysis: Suppose M is the total number of mesh points, uniformly distributed over p processors. Also, let t_{comp} be the time to update a mesh point without communication and let t_{comm} be the time

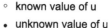

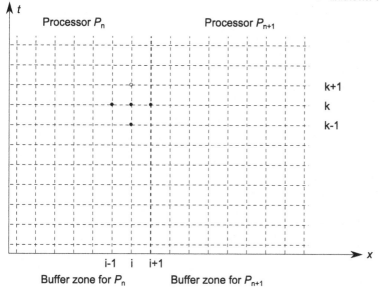

Fig. 7.3. Decomposition of computational domain into sub-domains for solving wave equations.

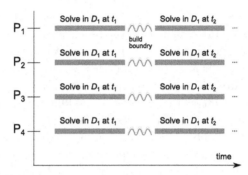

Fig. 7.4. PAT of decomposition of computational domain into sub-domains for solving wave equations.

to communicate one mesh point to a neighboring processor. The total time for one processor to solve such a problem is:

$$T(1, M) = M t_{\text{comp}} \tag{7.9}$$

With p processors, the time is:

$$T(p, M) = \frac{M}{p} t_{\text{comp}} + 2 t_{\text{comm}} \tag{7.10}$$

Thus, the speedup is:

$$S(p, M) = \frac{T(1, M)}{T(p, M)} = \frac{p}{1 + \frac{2p}{m} \frac{t_{\text{comm}}}{t_{\text{comp}}}} \tag{7.11}$$

and the overhead is:

$$h(p, M) = \frac{2p}{M} \frac{t_{\text{comm}}}{t_{\text{comp}}} \tag{7.12}$$

It is obvious that this algorithm is likely the

(a) The simplest parallel PDE algorithm.
(b) $h(P, N) \propto P/N$.
(c) $h(P, N) \propto t_c/t_0$.

It is quite common for algorithms for solving PDEs to have such overhead dependencies on granularity and communication-to-computation ratio.

We can easily observe the following:

(a) Overhead is proportional to $\frac{P}{M} = \frac{1}{m}$ where m is the number of mesh points on each processor. This means that the more mesh points each processor holds, the smaller the overhead.

(b) Overhead is proportional to $\frac{t_{\text{comm}}}{t_{\text{comp}}}$. This is a computer-inherent parameter.

2D Wave equation

Consider the following wave equation:

$$u_{tt} = c^2(u_{xx} + u_{yy}), \quad t > 0 \tag{7.13}$$

with proper BC and IC. It is very easy to solve this equation on a sequential computer, but on a parallel computer, it is a bit more complex. We perform a little difference operation (applying central differences on both sides):

$$\frac{u_{ij}^{k+1} + u_{ij}^{k-1} - 2u_{ij}^k}{\Delta t^2} - \frac{u_{i+1,j}^k + u_{i-1,j}^k - 2u_{ij}^k}{\Delta x^2} + \frac{u_{ij+1}^k + u_{i,j-1}^k - 2u_{ij}^k}{\Delta y^2}$$
$$\tag{7.14}$$

After further simplification,

$$u_{ij}^{k+1} = f(u_{ij}^{k-1}, u_{ij}^k, u_{i+1,j}^k, u_{i-1,j}^k, u_{i,j+1}^k, u_{i,j-1}^k) \tag{7.15}$$

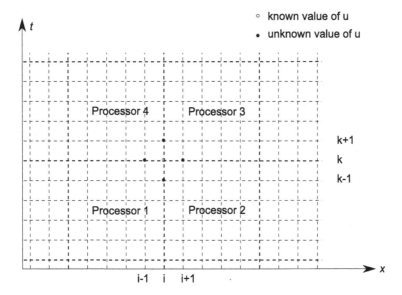

Fig. 7.5.   Decomposing the computational domain into sub-domains.

7.2.2. *3D Heat equation*

Consider the following 3D heat equation:

$$\begin{cases} u_t = u_{xx} + u_{yy} + u_{zz} \\ \text{Proper BC and IC} \end{cases} \tag{7.16}$$

After applying finite differences (forward-difference for the first-order time derivative and central-difference for the second-order spatial derivatives), we obtain:

$$\frac{u_{i,j,k}^{t+1} + u_{i,j,k}^t}{\Delta t} = \frac{u_{i+1,j,k}^t + u_{i-1,j,k}^t - 2u_{i,j,k}^t}{\Delta x^2} + \frac{u_{i,j+1,k}^t + u_{i,j-1,k}^t - 2u_{i,j,k}^t}{\Delta y^2}$$
$$+ \frac{u_{i,j,k+1}^t + u_{i,j,k-1}^t - 2u_{i,j,k}^t}{\Delta z^2} \tag{7.17}$$

Assuming $\Delta x = \Delta y = \Delta z = \Delta t = 1$, we get:

$$u_{i,j,k}^{t+1} = u_{i,j,k}^t + (u_{i+1,j,k}^t + u_{i-1,j,k}^t) + (u_{i,j+1,k}^t + u_{i,j-1,k}^t) + (u_{i,j,k+1}^t$$
$$+ u_{i,j,k-1}^t) - 6u_{i,j,k}^t \tag{7.18}$$

Therefore, solution of heat equation is equivalent to solving Poisson equation at every time step. In other words, we need to solve the Poisson equation (by iteration) at every time step.

7.2.3. *2D Poisson equation*

Consider the following 2D Poisson equation:

$$\begin{cases} u_{xx} + u_{yy} = f(x, y) \\ \text{Proper BC} \end{cases} \tag{7.19}$$

Applying central differences on both sides, we get:

$$\frac{u_{i+1,j} + u_{i-1,j} - 2u_{i,j}}{\Delta x^2} + \frac{u_{i,j+1} + u_{i,j-1} - 2u_{i,j}}{\Delta y^2} = f(x_i, y_j) \tag{7.20}$$

Assuming $\Delta x = \Delta y = 1$ (without losing generality), we can get the following simplified equation:

$$u_{i+1,j} + u_{i-1,j} + u_{i,j+1} + u_{i,j-1} - 4u_{i,j} = f(x_i, y_j) \tag{7.21}$$

We can have the following linear system of algebraic equations:

$$\begin{cases} u_{2,1} + u_{0,1} + u_{1,2} + u_{1,0} - 4u_{1,1} = f(x_1, y_1) \\ u_{3,2} + u_{1,2} + u_{2,3} + u_{2,1} - 4u_{2,2} = f(x_2, y_2) \\ \cdots \end{cases} \qquad (7.22)$$

If we define a new index,

$$m = (j-1)X + i \qquad (7.23)$$

to arrange a 2D index (i, j) to a 1D in $m = 1, 2, \ldots, X, X+1, X+2, \ldots, XY$. We can arrange the following 2D discrete Poisson equation:

$$\mathcal{A}u = B \qquad (7.24)$$

where $\mathcal{A}$ is a sparse matrix defined as:

$$\mathcal{A} = \begin{pmatrix} A_1 & I & \cdots & 0 & 0 \\ I & A_2 & \cdots & 0 & 0 \\ \vdots & \vdots & \ddots & \vdots & \vdots \\ 0 & 0 & \cdots & A_{Y-1} & I \\ 0 & 0 & \cdots & I & A_Y \end{pmatrix}_{Y \times Y}$$

This is a $Y \times Y$ matrix with sub-matrix elements

$$A_i = \begin{pmatrix} -4 & 1 & \cdots & 0 & 0 \\ 1 & -4 & \cdots & 0 & 0 \\ \vdots & \vdots & \ddots & \vdots & \vdots \\ 0 & 0 & \cdots & -4 & 1 \\ 0 & 0 & \cdots & 1 & -4 \end{pmatrix}_{X \times X}$$

which is a $X \times X$ matrix and there are Y of them, $i = 1, 2, \ldots, Y$. The final matrix $\mathcal{A}$ is a $(XY) \times (XY)$ matrix.

Therefore, solution of 2D Poisson equation is essentially a solution of a five-diagonal system.

7.2.4. *3D Poisson equation*

Consider the following 2D Poisson equation:

$$\begin{cases} u_{xx} + u_{yy} + u_{zz} = f(x, y, z) \\ \text{Proper BC} \end{cases} \qquad (7.25)$$

Applying central differences on both sides, we get:

$$\frac{u_{i+1,j,k} + u_{i-1,j,k} - 2u_{i,j,k}}{\Delta x^2} + \frac{u_{i,j+1,k} + u_{i,j-1,k} - 2u_{i,j,k}}{\Delta y^2}$$
$$+ \frac{u_{i,j,k+1} + u_{i,j,k-1} - 2u_{i,j,k}}{\Delta z^2} = f(x_i y_j z_k) \tag{7.26}$$

Assuming $\Delta x = \Delta y = \Delta z = 1$ (without losing generality), we can get the following simplified equation:

$$u_{i+1,j,k} + u_{i-1,j,k} + u_{i,j+1,k} + u_{i,j-1,k} + u_{i,j,k+1} + u_{i,j,k-1} - 6u_{i,j,k}$$
$$= f(x_i, y_j, z_k) \tag{7.27}$$

We can have the following linear system of algebraic equations:

$$\begin{cases} u_{2,1,1} + u_{0,1,1} + u_{1,2,1} + u_{1,0,1} + u_{1,1,2} + u_{1,1,0} - 6u_{1,1,1} = f(x_1, y_1, z_1) \\ u_{3,2,2} + u_{1,2,2} + u_{2,3,2} + u_{2,1,2} + u_{2,2,3} + u_{2,2,1} - 6u_{2,2,2} = f(x_2, y_2, z_2) \\ \cdots \end{cases}$$
$$\tag{7.28}$$

If we define a new index,

$$m = (k-1)XY + (j-1)X + i \tag{7.29}$$

We can linearize a 3D index (i, j, k) by a 1D index (m). Defining $u_m = u_{ijk}$, we obtain the following 3D discrete Poisson equation,

$$Au = b \tag{7.30}$$

where $\mathcal{A}$ is a sparse matrix defined as:

$$\mathcal{A} = \begin{pmatrix} A^1 & I & \cdots & 0 \\ I & A^2 & \cdots & 0 \\ \vdots & \vdots & \ddots & \vdots \\ 0 & 0 & \cdots & A^Z \end{pmatrix}_{Z \times Z}$$

whose sub-matrix elements are:

$$a^k = \begin{pmatrix} a_1^k & I & \cdots & 0 \\ I & a_2^k & \cdots & 0 \\ \vdots & \vdots & \ddots & \vdots \\ 0 & 0 & \cdots & a_y^k \end{pmatrix}_{Y \times Y}$$

And this is a $Y \times Y$ matrix with sub-matrix elements,

$$a_j^k = \begin{pmatrix} -6 & 1 & \cdots & 0 \\ 1 & -6 & \cdots & 0 \\ \vdots & \vdots & \ddots & \vdots \\ 0 & 0 & \cdots & -6 \end{pmatrix}_{X \times X}$$

Thus, $\mathcal{A}$ is a seven-diagonal $(XYZ) \times (XYZ)$ spare matrix and solution of the above equation requires an iterative method.

7.2.5. *3D Helmholtz equation*

Consider the following 3D Helmholtz equation:

$$\begin{cases} u_{xx} + u_{yy} + u_{zz} + \alpha^2 u = 0 \\ \text{Proper BC} \end{cases} \quad (7.31)$$

There are two families of methods in solving Helmholtz equation:

(1) Method of moments, aka, boundary element method.
(2) Method of finite differences.

In the following, we describe the finite-difference method:

$$\frac{u_{i+1,j,k} + u_{i-1,j,k} - 2u_{i,j,k}}{\Delta x^2} + \frac{u_{i,j+1,k} + u_{i,j-1,k} - 2u_{i,j,k}}{\Delta y^2}$$
$$+ \frac{u_{i,j,k+1} + u_{i,j,k-1} - 2u_{i,j,k}}{\Delta z^2} + \alpha^2 u_{i,j,k} \quad (7.32)$$

With the assumption that $\Delta x = \Delta y = \Delta z = 1$, we can simplify the above equation as:

$$u_{i+1,j,k} + u_{i-1,j,k} + u_{i,j+1,k} + u_{i,j-1,k} + u_{i,j,k+1} + u_{i,j,k-1}$$
$$+ (\alpha^2 - 6)u_{i,j,k} = 0 \quad (7.33)$$

Solution of this discrete equation is very similar to that of the discrete Poisson equation. Both methods require solution of sparse system of equations.

7.2.6. *Molecular dynamics*

Molecular dynamics is a popular approach for modeling many systems in biomolecule, chemistry and physics. Classical MD involves solutions of the following system of ODEs:

$$\begin{cases} m_i x_i'' = f_i(\{x_i\}, \{v_i\}) \\ x_i\,(t = 0) = a_i \qquad \forall\, i = 1, 2, \ldots, N \\ x_i'\,(t = 0) = v_i, \end{cases} \tag{7.34}$$

The key computing issues for MD are:

(1) MD can be highly nonlinear depending on the force fields. Most of the computing time is spent on force calculation.
(2) Force matrix defines the force from all pairs of particles or molecules,

$$F = \begin{pmatrix} f_{11} & \cdots & f_{1n} \\ \vdots & \ddots & \vdots \\ f_{n1} & \cdots & f_{nn} \end{pmatrix}$$

and it has the following important properties:

(a) Anti-symmetrical resulting from Newton's third law.
(b) Diagonal elements are all zero (no self-interaction).
(c) Sum of one complete row is the force acting on one particle by all others.
(d) Sum of one complete column is the force exerted by one particle on all others.
(e) The matrix is dense and complexity of obtaining the force is $O(N^2)$ if particles are involved in long-ranged interactions such as Coulomb's force.
(f) The matrix is sparse and complexity of obtaining the force is $O(N)$ if particles are involved in short-ranged interactions such as bound force only.

There are many ways to decompose the system. The two most popular ways are: Particle decomposition and spatial decomposition. A third way, not as common, considers force decomposition. Interaction ranges (short-, medium-, long-) determine the decomposition methods:

(1) **Particle decomposition:** Decomposing the system based on particle ID's regardless of particle positions. For example, for 100 particles

labeled as $(1, 2, 3, \ldots, 100)$ and 10 processors labeled as $P_1, \ldots, P_{10}$, we allocate 10 particles on each processor. Particles $(1, 2, \ldots, 10)$ are partitioned on P_1; $(11, 12, \ldots, 20)$ on $P_2, \ldots$, and $(91, 92, \ldots, 100)$ on P_{10}.

(2) **Spatial decomposition:** Decomposing the system based on particle positions regardless of particle ID's

   (i) Cartesian coordinates
   (ii) Cylindrical coordinates
   (iii) Spherical coordinates

For example, for 100 particles labeled as $(1, 2, \ldots, 100)$ and nine processors, we allocated approximately 11 particles on each processor. We divide the computational domain into $3 \times 3 = 9$ sub-domains. Particles $(1, 5, \ldots, 19)$ on P_1; $(12, 14, \ldots, 99)$ on P_2 and so on.

(3) **Force decomposition:** Decomposing the system based on force terms regardless of particle ID's and positions.

Of course, it is not uncommon; a mix of the above decomposition is used for one MD simulation.

Calculation of forces Typical force fields for protein simulation, for example, can be looked up from CHARM or AMBER library. Armies of professionals on physical chemistry and experimental structure biology provide the ever refined force formulation and the coefficients associated:

$$V = \sum_{\text{bonds}} k_b(b - b_0)^2 + \sum_{\text{angles}} k_\theta(\theta - \theta_0)^2 + \sum_{\text{dihedrals}} k_\phi[1 + \cos(n\phi - \delta)]$$

$$+ \sum_{\text{impropers}} k_\omega(\omega - \omega_0)^2 + \sum_{\text{Urey-Bradley}} k_u(u - u_0)^2$$

$$+ \sum_{\text{nonbonded}} \epsilon\left[\left(\frac{R_{\min_{ij}}}{r_{ij}}\right)^{12} - \left(\frac{R_{\min_{ij}}}{r_{ij}}\right)^6\right] + \frac{q_i q_j}{\epsilon r_{ij}} \tag{7.35}$$

Term-1 accounts for the bond stretches where k_b is the bond-force constant and $b - b_0$ is the distance from equilibrium that the atom has moved.

Term-2 accounts for the bond angles where k_θ is the angle-force constant and $\theta - \theta_0$ is the angle from equilibrium between three bounded atoms.

Term-3 is for the dihedrals (a.k.a. torsion angles) where k_ϕ is the dihedral force constant, n is the multiplicity of the function, ϕ is the dihedral angle and δ is the phase shift.

Term-4 accounts for the impropers, that is out of plane bending, where k_ω is the force constant and $\omega - \omega_0$ is the out of plane angle. The Urey-Bradley component (cross-term accounting for angle bending using 1, 3 non-bonded interactions) comprises Term-5 where k_u is the respective force constant and u is the distance between the 1, 3 atoms in the harmonic potential.

Non-bonded interactions between pars of atoms are represented by the last two terms. By definition, the non-bonded forces are only applied to atom pairs separated by at least three bonds. The van Der Waals (VDW) energy is calculated with a standard 12–6 Lennard-Jones potential and the electrostatic energy with a Coulombic potential.

The Lennard-Jones potential (Fig. 7.6) is a good approximation of many different types of short-ranged interaction besides modeling the VDW energy.

$$V_{LJ}(r) = 4\epsilon \left[\left(\frac{\sigma}{r} \right)^{12} - \left(\frac{\sigma}{r} \right)^{6} \right] \tag{7.36}$$

where ϵ is the depth of the depth of the potential well, σ is the (finite) distance at which the inter-particle potential is zero, and r is the distance between the particles. These parameters can be fitted to reproduce experimental data or mimic the accurate quantum chemistry calculations.

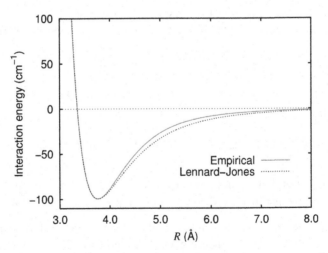

Fig. 7.6. Lennard-Jones potential, compared with empirical data.[1]

[1]R. A. Aziz, A highly accurate interatomic potential for argon, *J. Chem. Phys.* Vol. 99 Issue 6 (1993) 4518.

The r^{-12} term describes Pauli repulsion at short ranges due to overlapping electron orbitals and the r^{-6} term describes attraction at long ranges such as the van der Waals force or dispersion force.

The last term, the long-ranged force in the form of Coulomb potential between two charged particles,

$$\vec{F}_{ij} = \frac{q_i q_j}{|\vec{X}_i - \vec{X}_j|^3}(\vec{X}_i - \vec{X}_i) \tag{7.37}$$

The total force on particle i by all other particles is:

$$\vec{F}_i = \sum_{j \neq i}^{n} \frac{q_i q_j}{|\vec{X}_i - \vec{X}_j|^3}(\vec{X}_i - \vec{X}_i) \tag{7.38}$$

Estimating the forces cost 90% of the total MD computing time. But, solution of the equation of motion is of particular applied Mathematics interest as it involves solving the system of ODEs.

First, we may use the second order Verlet method:

$$a(t) = \frac{d^2 x}{dt^2} = \frac{x(t+\delta) + x(t-\delta) - 2x(t)}{\delta^2} \tag{7.39}$$

where $a(t)$ is the acceleration. So, the next time step can be calculated from the present and the previous steps by:

$$\begin{aligned} x(t+\delta) &= 2x(t) - x(t-\delta) + a(t)\delta^2 \\ &= 2x(t) - x(t-\delta) + \frac{F(t)}{m}\delta^2 \end{aligned} \tag{7.40}$$

Second, we may consider the Runge–Kutta method:

$$\begin{aligned} y' &= f(t,y) \\ y(t+h) &= y(t) + \frac{1}{6}h(k_1 + 2k_2 + 2k_3 + k4) \\ k_1 &= f(t,y) \\ k_2 &= f\left(t + \frac{1}{2}h, y + \frac{1}{2}k_1\right) \\ k_3 &= f\left(t + \frac{1}{2}h, y + \frac{1}{2}k_2\right) \\ k_4 &= f(t+h, y+k_3) \end{aligned} \tag{7.41}$$

It turns out this portion of the MD, i.e. actual solution of the system of ODEs, takes insignificant amount of computing time.

CHAPTER 8

FOURIER TRANSFORMS

The discrete Fourier transform (DFT) has an important role in many scientific and engineering applications. In the computer simulation of biological, physical and chemical phenomena, the calculation of long-range interaction depends heavily on the implementation of fast Fourier transform (FFT), which is a revolutionary FT algorithm that can compute the DFT of an n-point series in $\Theta(n \log n)$ time. Other areas including fluid dynamics, signal processing and time series are also utilizing FFT.

In this chapter, we discuss the simple form of serial FFT algorithm. Based on this, we then discuss the general parallel scheme for such algorithm. Finally, we introduce a high performance 3D FFT specially suited for scalable supercomputers.

8.1. Fourier Transforms

Mathematically, the following integrations define the Fourier transforms and its inverse:

$$g(\omega) = \frac{1}{\sqrt{2\pi}} \int_{-\infty}^{+\infty} f(t)e^{-i\omega t}dt \qquad (8.1)$$

$$f(t) = \frac{1}{\sqrt{2\pi}} \int_{-\infty}^{+\infty} g(\omega)e^{-i\omega t}d\omega \qquad (8.2)$$

Naturally, function $f(t)$ is defined in t-space (or physical space) and function $g(\omega)$ is defined in the k-space (or frequency space, or Fourier space). The key purpose of the transform is to study the way of representing general functions by sums of simpler trigonometric functions, as evident by Fig. 8.1.

Fourier transforms have broad applications including analysis of differential equations that can transform differential equations into algebraic equations, and audio, video and image signal processing.

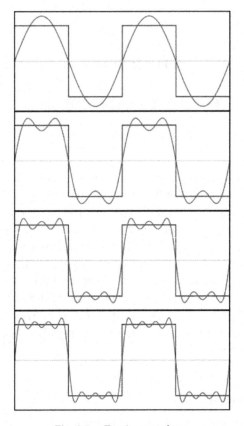

Fig. 8.1. Fourier transform.

8.2. Discrete Fourier Transforms

The sequence of N complex numbers $x_0, x_1, \ldots, x_{N-1}$ is transformed into the sequence of N complex numbers $X_0, X_1, \ldots, X_{N-1}$ by the DFT according to the formula:

$$X_k = \mathcal{F}(x) = \sum_{j=0}^{N-1} x_j e^{-2\pi \frac{ijk}{N}}, \quad \forall k = 0, 1, 2, \ldots, N-1 \qquad (8.3)$$

The inverse discrete Fourier transform (IDFT) is given by:

$$x_k = \mathcal{F}^{-1}(X) = \sum_{j=0}^{N-1} X_j e^{2\pi \frac{ijk}{N}}, \quad \forall j = 0, 1, 2, \ldots, N-1 \qquad (8.4)$$

It is obvious that the complex numbers X_k represent the amplitude and phase of the different sinusoidal components of the input "signal" x_n. The DFT computes the X_k from the x_n, while the IDFT shows how to compute the x_n as a sum of sinusoidal components $\frac{1}{N}X_k e^{2\pi\frac{ijk}{N}}$ with frequency k/N cycles per sample.

1D sequence can be generalized to a multidimensional array $x(n_1, n_2, \ldots, n_d)$ a function of d discrete variables,

$$n_l = 0, 1, \ldots, N_l - 1 \quad \forall l \in [1, d]$$

The DFT is defined by:

$$
X_{k_1, k_2, \ldots, k_d}
= \sum_{n_1=0}^{N_1-1}\left(\omega_{N_1}^{k_1 n_1} \sum_{n_2=0}^{N_2-1}\left(\omega_{N_2}^{k_2 n_2} \cdots \sum_{n_d=0}^{N_d-1} \omega_{n_d=0}^{k_d n_d} \cdot x_{n_1, n_2, \ldots, n_d} \right) \cdots \right) \quad (8.5)
$$

In real applications, 3D transforms are the most common.

8.3. Fast Fourier Transforms

FFT is an efficient algorithm to compute DFT and its inverse. There are many distinct FFT algorithms involving a wide range of mathematics, from simple complex-number arithmetic to group theory and number theory.

Computing DFT directly from the definition is often too slow to be practical. Computing a DFT of N points in the naïve way, using the definition, takes $O(N^2)$ arithmetical operations. An FFT can get the same result in only $O(N \log N)$ operations. The difference in speed can be substantial, especially for large data sets where $N = 10^3 \sim 10^6$, the computation time can be reduced by several orders of magnitude. Since the inverse DFT is the same as the DFT, but with the opposite sign in the exponent and a $1/N$ factor, any FFT algorithm can easily be adapted for it.

Simply put, the complexity of FFT can be estimated this way. One $FT(M \to M)$ can be converted to two shortened FTs as $FT(M/2 \to M/2)$ through change of running indices. If this conversion is done recursively, we can reduce the complexity from $O(M^2)$ to $O(M \log M)$.

Colloquially, FFT algorithms are so commonly employed to compute DFTs that the term "FFT" is often used to mean "DFT".

There are many FFT algorithms and the following is a short list:

(1) Split-radix FFT algorithm.
(2) Prime-factor FFT algorithm.
(3) Bruun's FFT algorithm.
(4) Rader's FFT algorithm.
(5) Bluestein's FFT algorithm.
(6) Butterfly diagram — a diagram used to describe FFTs.
(7) Odlyzko–Schönhage algorithm — applies the FFT to finite Dirichlet series.
(8) Overlap add/Overlap save — efficient convolution methods using FFT for long signals.
(9) Spectral music — involves application of FFT analysis to musical composition.
(10) Spectrum analyzers — Devices that perform an FFT.
(11) FFTW "Fastest Fourier Transform in the West" — C library for the discrete Fourier transform in one or more dimensions.
(12) FFTPACK — another C and Java FFT library.

Butterfly algorithm — the best-known FFT algorithm, is also known as the Cooley–Tukey algorithm.[1]

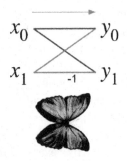

A radix-2 decimation-in-time FFT is the simplest and most common form of the Cooley–Tukey algorithm, although highly optimized Cooley–Tukey algorithm implementations typically use other form of the algorithm

---

[1] James W. Cooley was a programmer on John von Neumann's computer at the Institute for Advanced Study at Princeton (1953–1956) and retired from IBM in 1991.

as described below. Radix-2 DIT divides a DFT of size N into two interleaved DFTs of size $N/2$ with each recursive stage.

The discrete Fourier transform (DFT) is defined by the formula:

$$X_k = \sum_{n=0}^{N-1} x_n e^{-2\pi \frac{ink}{N}}, \tag{8.6}$$

where k is an integer ranging from to $N-1$.

Radix-2 DIT first computes the DFTs of the even-indexed x_{2m} $(x_0, x_2, \ldots, x_{N-2})$ inputs and of the odd-indexed inputs x_{2m+1} $(x_1, x_3, \ldots, x_{N-1})$ and then combines those two results to produce the DFT of the whole sequence.

More specifically, the Radix-2 DIT algorithm rearranges the DFT of the function x_n into two parts: a sum over the even-numbered indices $n = 2m$ and a sum over the odd-numbered indices $n = 2m + 1$:

$$X_k = \sum_{m=0}^{\frac{N}{2}-1} x_{2m} e^{-\frac{2\pi i}{N}(2m)k} + \sum_{m=0}^{\frac{N}{2}-1} x_{2m+1} e^{-\frac{2\pi i}{N}(2m+1)k} \tag{8.7}$$

Thus, the whole DFT can be calculated as follows:

$$X_k = \begin{cases} E_k + e^{-\frac{2\pi i}{N}k} O_k, & \text{if } k < N/2 \\[2mm] E_{k-N/2} - e^{-\frac{2\pi i}{N}(k-N/2)} O_{k-N/2}, & \text{if } k \geq N/2 \end{cases} \tag{8.8}$$

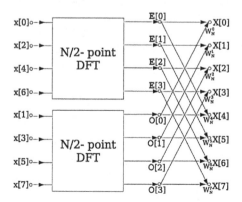

Another graph depicting the FFT algorithm:

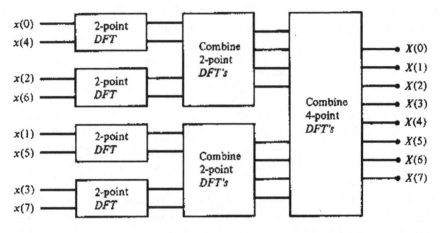

First, we divide the transform into odd and even parts (assuming M is even):

$$X_k = E_k + \exp\frac{2\pi ik}{M}O_k$$

and

$$X_{k+M/2} = E_k - \exp\frac{2\pi ik}{M}O_k$$

Next, we recursively transform E_k and O_k to the next two 2×2 terms. We iterative continuously as shown in Fig. 8.2 until we only have one point left for Fourier transform. In fact, one point does not need any transform.

```
Function Y=FFT(X,n)
if (n==1)
     Y=X
else
     E=FFT({X[0],X[2],...,X[n-2]},n/2)
     O=FFT({X[1],X[3],...,X[n-1]},n/2)
     for j=0 to n-1
          Y[j]=E[j mod (n/2)]
               +exp(-2*PI*i*j/(n/2))*O[j mod (n/2)]
     end for
end if
```

Fig. 8.2.   Pseudo code for serial FFT.

General case:

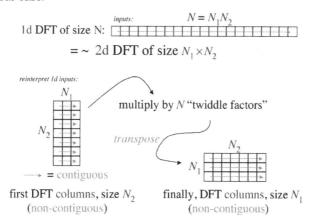

Cooley–Tukey algorithms recursively re-express a DFT of a composite size N = N1N2 as the following:

- Perform N1 DFTs of size N2.
- Multiply by complex roots of unity.
- Perform N2 DFTs of size N1.

8.4. Simple Parallelization

Based on the serial FFT algorithm, a nature implementation of parallel FFT is to divide the task according to the recursive algorithm until all the processors has its own stake of subtasks. By doing this, every processor runs independently on its own subtasks and communication occurs only when merging back the results.

Consider a simple example of computing 16-point FFT on four processors. The data can be partitioned as in Fig. 8.3 so the calculations of 4-point FFT are entirely local. In the second stage, the calculation of even and odd subsequence requires P_0 to communicate with P_2 and P_1 with P_3. In the final stage, P_0 and P_1 communicate and so do P_2 and P_3.

This algorithm, although simple and closely aligned with the original serial algorithm, is in fact quite efficient especially in the cases where the data points are far more than the number of processors. The communication happens $\log P$ times in the whole calculation and in each communication, every processor sends and receives N/P amount of data.

P_0	P_1	P_2	P_3
0	1	2	3
4	5	6	7
8	9	10	11
12	13	14	15

Fig. 8.3.   The scatter partition for performing 16-point FFT on four processors.

8.5. The Transpose Method

The nature improvement from the above simple parallel algorithm is the so-called transpose method that eliminates the $\log P$ communications to one all-to-all transpose communication. To understand how this works, it is necessary to have a deeper insight into the FFT algorithm.

As a recursive algorithm, the original FFT can always be written in the form of a forward loop that involves exactly $\log_2 N$ steps. Moreover, the total number of data is unchanged after every merge. Thus, the data points X_i at step.

Take the example of the 16-point FFT on four processors in the previous section.

One $FT(M \times M)$ can be converted into two shortened FTs as FT $(\frac{M}{2} \times \frac{M}{2})$ through a change of running variables. If this conversion is done recursively, the complexity can be reduced from $O(M^2)$ to $O(M \log M)$. This new transform is the FFT. Here are the details.

First, divide the transform into odd and even parts (assuming M is even):

We then have:

$$X_k = E_k + \exp\left(2\pi \frac{ik}{M}\right) O_k, \ k = 0, 1, \ldots, M' \tag{8.9}$$

$$X_{k+\frac{M}{2}} = E_k - \exp\left(2\pi \frac{ik}{M}\right) O_k, \ k = 0, 1, \ldots, M' \tag{8.10}$$

Next, transform E_k and O_k recursively to the next 2×2 terms. We iterate continuously until we only have one point left for Fourier Transform. In fact, one point does not need any transform.

8.6. Complexity Analysis for FFT

Define the following terms:

$$T(M) = \text{Time for M point transform} \tag{8.11}$$

$$T_{\text{comp}} = \text{Time for real number, X + or } - \tag{8.12}$$

$$T_{\text{mult}} = \text{Time for multiplying 2 complex numbers} = 6T_{\text{comp}} \tag{8.13}$$

$$T_{\text{pm}} = \text{Time for adding two complex} \tag{8.14}$$

Therefore, using the formulae above, we get the following relationship:

$$T(M) = 2T\left(\frac{M}{2}\right) + M\left(T_{\text{pm}} + \frac{T_{\text{mult}}}{2}\right) \tag{8.15}$$

Iteratively, we have:

$$T(M) = \left(T_{\text{pm}} + \frac{T_{\text{mult}}}{2}\right) M \log M \tag{8.16}$$

Let us define:

$$T_+ = 2T_{\text{pm}} + T_{\text{mult}} \tag{8.17}$$

Thus obtaining the formula for the single processor time as:

$$T(1, M) = T_+ M \log M \tag{8.18}$$

Now, Let us discuss the details of FT parallelization.

For simple FFT: We can decompose the task by physical space, fourier space, or both. In any of these cases, we can always obtain perfect speed up, as this problem is the EP problem.

For Complete FFT ($M \to M$ on P processors): We assume M and P are factors of $2N$ and $m = \frac{M}{P}$ is an integer.

Case I $M > P$:

$$T(P, N) = 2T\left(P, \frac{M}{2}\right) + \frac{T_+}{2}\frac{M}{P} \tag{8.19}$$

$$= \left(\frac{M}{P}\right)T(P, M) + \frac{T_+}{2}\log\left(\frac{M}{P}\right) \tag{8.20}$$

Case I $M \leq P$:

$$T(P, M) = T\left(P, \frac{M}{2}\right) + T_- = T_- \log M \tag{8.21}$$

Substituting this equation into the above, we get:

$$T(P, M) = \frac{M}{2p}\left[2T_- \log P + T_+ \log \frac{M}{p}\right] \tag{8.22}$$

$$= \frac{M}{2P}[T_+ \log M + (2T_- - T_+) \log P] \tag{8.23}$$

Therefore, the speedup is:

$$S(P, M) = \frac{P}{1 + \frac{2T_- - T_+}{T_+}\frac{\log P}{\log M}} \tag{8.24}$$

And the overhead is:

$$h(P, M) = \frac{2T_- - T_+}{T_+}\frac{\log P}{\log M} \tag{8.25}$$

We can estimate T_- and T_+ in terms of the traditional T_{comp} and T_{comm}:

$$T_+ = 2T_{pm} + T_{mult} \approx 10T_{comp} \tag{8.26}$$

$$T_- = T_{pm} + T_{mult} + T_{shift} = 8T_{comp} + 2T_{comm} \tag{8.27}$$

Therefore, the overhead is:

$$h(P, M) = \frac{2T_{comm} + T_{comp}}{5T_{comp}}\frac{\log P}{\log M} \tag{8.28}$$

Remarks

1. Communication to computation ratio affects overhead $h(P, M)$.
2. $\frac{P}{M}$ affects overhood, too, but in a much smaller way than most other cases. For parallel FFT, overhead is typically very small.

CHAPTER 9

OPTIMIZATION

Mathematical optimization, aka, mathematical programming, is the selection of a best element from some sets of available alternatives with regard to some constraints. More generally, it involves finding best available values of some objective function (O.F.) given a defined domain. There are a variety of different types of O.F.'s and different types of domains, a fact that enriches optimization with many subfields: convex programming, integer programming, quadratic programming, nonlinear programming, and dynamic programming, as well as stochastic programming in which the parameters and constraints involve random variables. It is a rich field with theories originating from, and applications to, many disciplines including mathematics, science, finance, operations research, and engineering. It has been adopted by, arguably, the widest communities among these areas.

Finding the best available values is a techy venture, most commonly, of computational techniques. Such techniques include optimization algorithms (simplex algorithms by Dantzig, combinatorial algorithms, etc), heuristics, and iterative methods introduced by giants like Newton and Gauss. Monte Carlo methods are examples of such iterative methods relying on repeated random sampling to reach the "best available values". The MC methods are useful for finding the approximate solutions of problems, usually, with many coupled degrees of freedom, when their exact solutions are infeasible to find.

The commonality of these optimization problems is that vast amount of computing resources is required, which results in another commonality: parallel computing is the only way to obtain solutions, most commonly approximate solutions, to optimization problems. Unfortunately, there are still no widely accepted scalable parallel algorithms for optimizations. Indeed, this is a fertile field where challenges and opportunities co-exist.

Achieving scalable performance for MC methods on large parallel computers with thousands of processing cores is a serious challenge. There are still no agreeable "standard" parallel algorithms for MC methods. We will introduce in this manuscript the early stages of a few promising methods.

9.1. Monte Carlo Methods

Monte Carlo methods, as of today, still do not have a widely accepted definition. Regardless, Monte Carlo methods are a class of numerical algorithms that rely on repeated random sampling to compute their results. This much is clear: (1) it is a computational algorithm; (2) it requires massive repetitions; and (3) it is random sampling.

As mentioned earlier, the Monte Carlo methods are most suited for problems whose solutions are infeasible to achieve by deterministic approaches. They were first practiced in 1940s by von Neumann and his fellow Los Alamos scientists for the Manhattan Project. They are the only viable approaches for numerical integrations in high dimensions and with complicated domains. For example, for calculations in lattice gauge theory, the integration dimension can be as high as a million or higher. As one of the random walk Markov chain Monte Carlo methods, it can drive the chain of states, iteratively, to the region of optimality in search space.

Monte Carlo methods are, arguably, the most powerful numerical methods because it can solve problems that many others fail and it can solve many problems in broad diverse areas. However, Monte Carlo methods have serious deficiencies. First, the sampling process is long and the stop condition when the iteration is converged is usually not known a prior and, thus, most MC simulations are extremely computational time consuming although many attempts, such as the importance sampling for generating "more important samples" to reduce the variance from the desired distributions, were introduced to speed things up. Second, the iterative nature of the sampling process makes it difficult for parallelization, a promising remedy of the first. Yet, parallelization of the MC methods may enable solutions of many problems in engineering, science, and finance, as well as social sciences, leading breakthroughs in these and other fields.

9.1.1. *Basics*

The key components of typical Monte Carlo methods include:

(1) Probability distribution functions: This PDF describes the desired properties of the physical or mathematical systems being simulated.
(2) Random number generator. A source of random numbers uniformly distributed on the unit interval must be available and these flat

random numbers can be converted conveniently to other distributions as needed.

(3) Sampling rule: a prescription for sampling from the specified probability distribution function, assuming the availability of random numbers on the unit interval.

(4) Scoring: the outcomes must be accumulated into overall tallies or scores for the quantities of interest.

(5) Error estimation: an estimate of the statistical error, i.e., variance, as a function of the number of trials and other quantities must be determined.

(6) Variance reduction techniques: methods for reducing the variance in the estimated solution to reduce the computational time for Monte Carlo simulations.

The most important function of Monte Carlo methods is the ability to locate the global optimum while performing optimization. The objective functions of most scientific or engineering applications usually have multiple local minima (as in Fig. 9.1) and the ability to escape from such local minima is the greatest advantage of the Monte Carlo methods.

9.1.2. *Metropolis-Hastings Algorithm*

The Metropolis-Hastings algorithm, introduced by Metropolis and four colleagues in 1953 and extended by Hastings in 1970, is one of the Markov

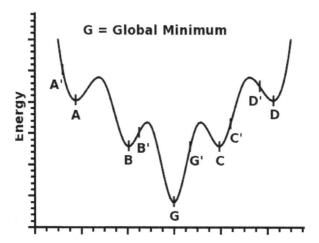

Fig. 9.1. A typical higher-dimensional Objectie Function (OF) "folder" to 1D.

chain Monte Carlo (MCMC) methods for obtaining a sequence of random samples from a probability distribution for which direct sampling is difficult. This sequence can be used to approximate the distribution or to perform other measurements in the generated distribution. This chain can also lead the search to a region of optimality.

Major steps of numerical Metropolis-Hastingsalgorithm include:

Step 0: Initiate a state X and compute its scalar objective function: $E(X)$;
Step 1: Make a small move from X to a new state $Y = X + \Delta X$;
Step 2: Compute the new objective function $E(Y)$ for the new state Y;
Step 3: Accept the new state Y according to the following probability

$$P(Y) = \min\left\{1, \exp\left(-\frac{E(Y) - E(X)}{T}\right)\right\} \qquad (9.1)$$

Step 4: If stop conditions fail, then go to Step 1;
Step 5: End.

In the Step 3 above, the acceptance probability depends on the difference of the O.F.'s at two nearby states X and Y and a parameter T yet to be introduced. This new parameter T, a positive real number, is widely regarded as temperature. It is, in fact, a parameter for regulating the level of forgiveness when accepting a state with higher O.F.

Examining the formula in Step 3, we notice

If $E(Y) < E(X)$, then $P(Y) = 1$ and the new state Y is always accepted.
If $E(Y) > E(X)$

If T is big, then $\frac{E(Y)-E(x)}{T}$ is small and $\exp(-\frac{E(Y)-E(x)}{T}) \to 1^-$ and thus a new bad state Y is more likely accepted.
(A special case: $T \to \infty \Rightarrow P(Y) \to 1$. Accepting any bad state).
If T is small, then $\frac{E(Y)-E(x)}{T}$ is big and $\exp(-\frac{E(Y)-E(x)}{T}) \to 0^+$ and thus a new bad state Y is more likely rejected.
(A special case: $T \to 0 \Rightarrow P(Y) \to 0$. Rejecting any bad state).

Therefore, it is not difficult to conclude that the parameter T is a tolerance for bad states. This tolerance naturally dictates the convergence rate of the simulation.

In the Metropolis-Hastings algorithm, it's a fixed number while dynamically adjusting it, an act called temperature scheduling, to achieve optimal convergence is the true state-of-the-art for this algorithm. As of today, there is no universal temperature schedule and it may never be possible to find one because an optimal temperature schedule depends on applications.

9.1.3. *Simulated annealing*

As an adaptation to the Metropolis-Hastings algorithm, simulate annealing, introduced by Kirkpatrick and two colleagues in 1983, also performs the following new state-generating step, i.e., the probability of accepting a new state Y from the old state X:

$$P(Y) = \min\left\{1, \exp\left(-\frac{E(Y) - E(x)}{T}\right)\right\} \tag{9.2}$$

The only difference is that the parameter T varies during the process or sampling. The precise form of change of the parameter T is still elusive but some basic principles should be followed or a poor choice of the parameter T may slow down the convergence or lead to diverge. Another rule of thumb is to allow more bad states to be accepted at the start of iterations than the latter stages, i. e., scheduling the parameter T to decrease as iteration progresses.

9.1.4. *Genetic algorithm*

A genetic algorithmis a search heuristic that mimics the process of natural evolution. This heuristic is used to generate useful solutions to optimizations and search problems. Genetic algorithms belong to the larger class of evolutionary algorithms inspired also by natural evolution, such as inheritance, mutation, selection, crossover, etc.

9.2. Parallelization

9.2.1. *Basics*

There are many approaches for parallelizing Monte Carlo methods and none appears ideal. The correct measure of success is still that of the shortest time for achieving the results. Many other claims of success include keeping all processors busy all the time or achieving a zillion Flops are common deceptions for parallel computing and such deceptions are particularly difficult to detect for Monte Carlo simulations.

The latest approaches of parallelizing Monte Carlo methods always involve decomposition of new state generation, or the search space, or the

Markov chains. The latter two are related and none appears to be able to scale for parallel computers with many processors.

9.2.2. *Chain mixing method*

With very limited numerical experiments and fantastic performances, I would introduce a new method: the chain mixing method.

The following are the basic steps:

Step 0: Start P independent Markov chains, one per processors, by Metropolis-Hastings algorithm or by the simulated annealing;

Step 1: All P processors elongate their individual Markov chains with individual random number sequences by Metropolis-Hastings algorithm or by the simulated annealing;

Step 2: At a given mixing time Ω, suspend chain elongation and evaluate the current states of all P chains and identify the best state;

Step 3: If the best state fails to satisfy stop conditions, one or more best state(s) are selected, according to the acceptance probability, and planted to P processors and go to Step 1;

Step 4: End.

As shown in Fig. 9.2, N processors start N Markov chains with initial states $X_1, X_2, \ldots, X_n$ and iterate until they reach states $Y_1, Y_2, \ldots, Y_n$ when a mixing evaluation occurs. At this time, all states except one best state Y_2 are terminated. All processors, including the one that generated state Y_2, are given the state of Y_2 and use their own random number sequences to elongate the Markov chains until next mixing, etc. This approach is efficient only for parallel machines with few processors.

An improvement of the above method is presented in Fig. 9.3. The difference is that two or more states are selected to "re-produce". We pool the ending states $Y_1, Y_2, \ldots, Y_n$ and compute the "average" O.F. of these states by

$$\overline{E} = \frac{E(Y_1) + E(Y_2) + \cdots + E(Y_n)}{n}$$

and then compare the acceptance factors of the ending states

$$P(Y_i) = \exp\left(-\frac{E(Y_i) - \overline{E}}{T}\right)$$

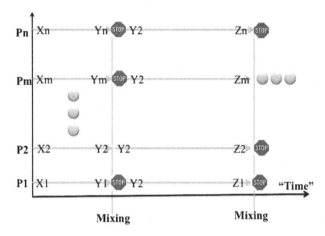

Fig. 9.2.   Mixing method 1 in which only one chain is selected to propagate.

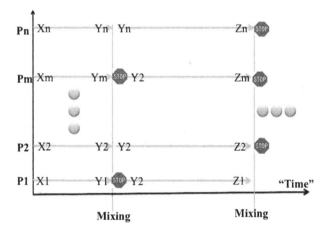

Fig. 9.3.   Mixing method 2 in which some chains are selected to propagate.

to assess re-production allowance. The processors with bad states will suspend own chain production and adopt others processors' better states to produce new Markov chains.

This schedule is a faithful parallelization of the Metropolis-Hastings algorithm or simulated annealing.

Parallelization of Monte Carlo methods is a serious challenge with great benefits. For the schemes we discussed here, we still have many parameters to select, e.g., the mixing periods and reproduction probability.

CHAPTER 10

APPLICATIONS

This chapter focuses on applications of the algorithms to physics and engineering. They include classical and quantum molecular dynamics, and Monte Carlo methods.

First-principle, parameter-free simulations are always desirable, but are rarely feasible for a large class of problems in physics and engineering. Unfortunately, computers available today and even decades to come will still be insufficient for most realistic problems.

Thus, second-principle modeling with some free parameters and simplifications of the original equations remains as the dominant approach in scientific computing. The essence of this approach rests on the representation of the physical systems by mathematical formulations. Scientists in each of their own areas are responsible for the conversion of physics to mathematics.

After the mathematical formulations are obtained, the immediate concern falls to the discretization of the equations for computer representation, which is largely, applied mathematicians' responsibilities.

The mathematical formulations are commonly divided into two major categories: Differential equations and global optimizations. For the former, we can get partial differential equations (PDEs) which are usually nonlinear and ordinary differential equations (ODEs) (which are, in many cases, stiff). These equations are largely solved by finite-different, finite-element, or particle method or their hybridization. While for the latter, we always get problems with multiple local minima, for which Monte Carlo methods usually prove to be advantageous.

The following is a short list of grand challenges covering problems of intense complexity in science, engineering, and technology. Only the highest performance computing platforms, i.e. parallel computing, may offer limited insights to these problems.

Parallel computing may bring revolutionary advances to the study of the following areas, which usually require large-scale computations. Scientists and engineers in these areas must seize the opportunity to make efficient use of this cost-effective and enabling technology.

Traditional methods such as the finite difference, finite element, Monte Carlo method, and particle methods, are still the essential ingredients for solving these problems except that these algorithms must be parallelized.

Fluid dynamics, turbulence, differential equations, numerical analysis, global optimization, and numerical linear algebra are a few popular areas of research.

Turbulence: Turbulence in fluid flows affects the stability and control, thermal characteristics, and fuel performance of virtually all aerospace vehicles. Understanding the fundamental physics of turbulence is requisite to reliably modeling flow turbulence for the analysis of realistic vehicle configuration.

Topics include simulation of devices and circuits, VLSI, and artificial intelligence (speech, vision, etc.).

Speech: Speech research is aimed at providing a communications interface with computers based on spoken language. Automatic speech understanding by computer is a large modeling and search problem in which billions of computations are required to evaluate the many possibilities of what a person might have said within a particular context.

Vision: The challenge is to develop human-level visual capabilities for computers and robots. Machine vision requires image signal processing, texture and color modeling, geometric processing and reasoning, and object modeling. A component vision system will likely involve the integration of all of these processes with close coupling.

10.1. Newton's Equation and Molecular Dynamics

This is the fundamental equation by Sir Isaac Newton (1642–1727) in classical mechanics, relating the force acting on a body to the change in its momentum over time. It is used in all calculations involving moving bodies in astrophysics, engineering, and molecular systems.

$$F = ma + \frac{dm}{dt}v \tag{10.1}$$

In molecular dynamics and N-body problems, or any models that involve "particles",

$$\left\{ m_i \frac{d^2x}{dt^2} = f_i(x_1, x_2, \ldots, x_N) \right\} \quad \forall\, i \leq N \tag{10.2}$$

Main issues are:

- Interaction range: short-, medium-, and long-.
- Force estimation (accounts for 90% of CPU usage).
- Solution of coupled Newton Equations (ODE system).
- Science (of course!).

The force matrix is given by,

$$F = \begin{pmatrix} f_{11} & f_{12} & \cdots & f_{1N} \\ f_{21} & f_{22} & \cdots & f_{2N} \\ \vdots & \vdots & \ddots & \vdots \\ f_{N1} & f_{N2} & \cdots & f_{NN} \end{pmatrix} \tag{10.3}$$

Remarks

- $N \times N$ particle interactions form a force matrix.
- Diagonal elements are self-interaction and are always zero.
- If the particles are named properly, F decreases as $|i - j|$ increases.
- The sum of the elements of row i is equal to the total force on particle i.
- Column j represents the forces on all particles by particle j; this is rarely useful.
- In a long-ranged context with N bodies, F is dense and has complexity $O(N^2)$.
- In a short-ranged context (MD), F is dense and has complexity $O(N)$.

Solution of the motion equation:

- 2nd order Valet method.
- 2nd order Runge–Kutta method.
- 4th order Runge–Kutta method.

Depending on the level of resolution required, one could apply 2nd, 4th, or 6th order ODE solvers.

10.1.1. *Molecular dynamics*

There are two types of molecular dynamics (MD) approaches currently practised by researchers in physics, chemistry, engineering, particularly in materials science. One is the first-principle approach by directly solving a N body system governed by the Schrödinger equation. This approach is called the quantum molecular dynamics (QMD) which considers all quantum mechanical effects. The other involves solving systems of Newton's equations governing particles moving under quasi-potentials. This approach is called classical molecular dynamics (CMD).

Both methods have their advantages and shortcomings. QMD considers full quantum mechanics and is guaranteed to be correct if the system contains a sufficient number of particles, which is infeasible with today's computing power. The largest system considered so far is less than 500 particles. CMD can include many more particles, but the potential used here is basically an averaging quantity and may be incorrectly proposed. If that is wrong, the whole results are wrong. So a study with multiple free parameters can make it risky.

CMD problems are usually solved by the following methods:

(1) Particle-particle (PP).
(2) Particle-mesh (PM).
(3) Multipole method.

In the PP method, the forces are computed exactly by the inter-particle distances, while in the PM method, the forces, being treated as a field quantity, are approximated on meshes. In multi-pole method, the field is expanded into multi-poles; the higher order poles, contributing negligibly to the field, are truncated.

There are two types of interactions measured by distance: Short-ranged and long-ranged. For short-ranged interactions, PP is a method of choice as the complexity is $O(N)$ where N is the number of particles. For long-ranged interactions, PP may be used, but is surely a bad choice as the complexity becomes $O(N^2)$.

For short-ranged interactions, the recorded value for $N = 108$ in 1994. If the same method is used for long-ranged interactions, $N = 104$, which is basically useless for realistic problems. Thus, a good choice of method for long-ranged interactions is the PM. The multi-pole method is unique for long-ranged interactions problems.

Molecular dynamics may be unique in offering a general computational tool for the understanding of many problems in physics and engineering. Although it is not a new idea, its potential has never been realized due to the lack of adequate computing source.

10.1.2. *Basics of classical MD*

In this section, we plan to,

(1) Develop general-purpose parallel molecular dynamics (MD) methodologies and packages on distributed-memory MIMD computers.

(2) Apply the resulting package to DNA-protein interaction analysis, thin-film depositions, and semi-conductor chip design as well as material design and analysis.

(3) Study microscopic interaction form (physics) and explore the novel macro properties in cases where interacting forms are known (engineering).

While molecular dynamics is a powerful method for many systems involving dynamics, continuum modeling without treating the system as an assemble of particles has been playing a major role for simulating fluid, gas, and solids. The major advantage of the continuum modeling is the dramatic reduction in computing time enabled by treat the system as a finite number of interacting blocks whose motion is governed by, usually, basic principles of energy and momentum conversations. Each of these blocks, in effect, contains thousands or tens of thousands or more particles, the basic entity in the MD simulations. Thus, continuum modeling can be, depending on size and nature of the system being simulated, thousands to millions of times faster than MD. However, the spatial and temporal resolutions offered by MD can be a feature that continuum modeling may never provide. It is important to select the right method for the right problem and, rather commonly, both methods may have to be used for one system for optimal accuracy and efficiency.

Molecular dynamics (MD) is in essence an N-body problem: Classical mechanical or quantum mechanical, depending on the scale of the problem. MD is widely used for simulating molecular-scale models of matter such as solid state physics, semiconductor, astrophysics, and macro-biology. In CMD, we consider solving N-body problems governed by Newton's second law, while for quantum MD, we solve systems governed by Schrödinger's equation.

Mathematically, MD involves solving a system of non-linear ordinary differential equation (ODE). An analytical solution of an N-body problem is hard to obtain and, most often, impossible to get. It is also very time-consuming to solve non-linear ODE systems numerically. There are two major steps in solving MD problems: Computing the forces exerted on each particle and solving the ODE system. The complexity lies mostly in the computation of force terms, while the solution of ODEs takes a negligible ($\sim 5\%$ of total) amount of time. There exist many efficient and accurate numerical methods for solving the ODEs: For example, Runge–Kutta, Leapfrong, and Predictor-corrector.

The computation of the force terms attracts much of the attention in solving MD problems. There are two types of problems in CMD: Long-ranged and short-ranged interactions. In long-range interactions, the complexity for a system with N particles is $O(N^2)$ and it seems to be hard to find a good method to improve this bound except to let every particle interact with every other particle. While for short range interaction, the complexity is $C \times N$ where C is a constant dependant on the range of interaction and is the method used to compute the forces.

The interests lie in measuring the macro physical quantities after obtaining the micro physical variables such as positions, velocities, and forces.

Normally, a realistic MD system may contain $N > 10^5$ particles, which makes it impossible for most computers to solve. As of summer 1993, the largest number of particles that have been attempted was $= 10^7$. This was done on the 512-processor Delta computer by Sandia National Laboratories researchers. Of course, the study has not been applied to real physics problems.

The cost is so high that the only hope to solve these problems must lie in parallel computing. So, designing an efficient and scalable parallel MD algorithm is of great concern. The MD problems that we are dealing with are classical N-body problems, i.e. to solve the following generally non-linear ordinary differential equations (ODE),

$$m_i \frac{d^2 x_i}{dt^2} = \sum_j f_{ij}(x_i, x_j) + \sum_{j,k} g_{ijk}(x_i, x_j, x_k) + \cdots , \quad I = 1, 2, \ldots, N$$

(10.4)

where m_i is the mass of particle i, x_i is its position, $f_{ij}(\cdot, \cdot)$ is a two body force, and $g_{ijk}(\cdot, \cdot, \cdot)$ is a three-body force. The boundary conditions and initial conditions are properly given. To make our study simple, we only consider two-force interactions, so $g_{ijk} = 0$.

The solution vector X, of the system can be written in the following iterative form:

$$X_{\text{new}} = X_{\text{old}} + \Delta \left(X_{\text{old}}, \frac{dX_{\text{old}}}{dt}, \ldots \right)$$

(10.5)

The entire problem is reduced to computing

$$\Delta \left(X_{\text{old}}, \frac{dX_{\text{old}}}{dt}, \ldots \right)$$

(10.6)

In fact the core of the calculation is that of the force. Typically, solving the above equation costs less than 5% of total time if the force terms are known. Normally, Runge–Kutta, or Predictor-Corrector, or Leapfrog methods are used for this purpose. More than 95% of the time is spent on computing the force terms. So we concentrate our parallel algorithm design on the force calculation.

There are two types of interactions: Long range and short-ranged forces. The long-ranged interactions occur often in gases, while short-ranged interactions are common in solids and liquids. For long-ranged forces, the cost is typically $O(N^2)$ and there are not many choices of schemes. However, for short range forces, the cost is generally $O(N)$ and there exists a variety of methods.

A force matrix (Fig. 10.1) is helpful in understanding the force calculation. The matrix element f_{ij} is the force acting on particle i by particle j. Thus, if we add together all elements in a row (say, row i), we get the total force on particle i. The matrix has several simple properties:

(1) The diagonal elements are zero.
(2) It is symmetric.
(3) The matrix can be sparse or dense depending on the system interaction range. For short-ranged interactions, it is sparse. For long-ranged interactions, it is dense.

Long-ranged interactions: Most N-body problems and study of plasma under Coulomb's interactions must consider long-ranged interactions.

• Method I: Particle-mesh method.
• Method II: Multi-pole method.
• Method III: The fast multi-pole method.
• Method IV: Rotation scheme.

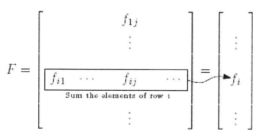

Fig. 10.1. The force matrix and force vector. The force matrix element f_j is the force on particle i exerted by particle j. Adding up the elements in row i, we get to total force acting on particle i by all other particles.

The total two-body force on particle i is given by,

$$F_i = \sum_{j \neq i} f_{ij} \tag{10.7}$$

So, if N particles are distributed uniformly to p processors (assuming $n = N/p$ is an integer), every particle must be considered by every processor for us to compute the total force.

As shown in Fig. 10.2, all processors simultaneously pass their own particles to their right-hand neighbors (RHN). Each processor will keep a copy of its own particles. As soon as RHN get their left-hand neighbor's particles, compute the partial forces exerted on the local particles by the incoming ones. Next time, all processors will "throw" out the incoming particles to their RHN. These RHN will again use the newly arrived particles to compute the remaining force terms for local particles. This process is repeated until all particles visit all processor.

Performance Analysis: First, we define some parameters. Let t_{xchg} be the time to move on particles from one processor to its neighbor and let t_{pair} be the time to compute the force for a pair of particles. We have that,

$$T(1, N) = N^2 t_{\text{pair}} \tag{10.8}$$

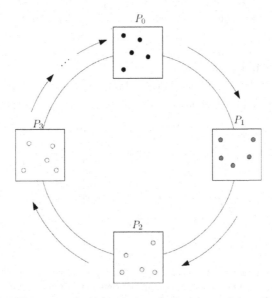

Fig. 10.2.   The communication structure for long-range interaction in MD.

and

$$T(p, N) = p(N^2 t_{\text{pair}} + N t_{\text{xchg}}) \tag{10.9}$$

Therefore, the speedup is:

$$S(p, N) = p \frac{1}{1 + \frac{1}{n} \frac{t_{\text{xchg}}}{t_{\text{pair}}}} \tag{10.10}$$

and the overhead is:

$$h(p, N) = \frac{1}{n} \frac{t_{\text{xchg}}}{t_{\text{pair}}} \tag{10.11}$$

Remarks

- $\frac{1}{n}$ is important for reducing overhead.
- $\frac{t_{\text{xchg}}}{t_{\text{pair}}}$ again controls the overhead.
- For long range interaction, $h(p, N)$ is typically very small.

Short-ranged interactions: For short-ranged interaction, the algorithm is very different. Basically, two major techniques (applicable to both serial or parallel) are widely used to avoid computing force terms that have negligible contribution to the total force: (1) particle-in-cell and (2) neighbor lists.

Assume that particles with a distance less than r_c have a finite, non-zero interaction, while particles with distances larger than r_c do not interact.

Particle-in-cell: Create a 3D cubic mesh in the computational domain with mesh size r_c. Therefore, particles in a certain cell of volume r^c only interact with particles in the 26 other adjacent cells. Thus, a volume of $27r^c$ of particles must be considered. It is not the best idea, but much better than considering the entire domain (as in long interaction case) which typically contains $O(10^3)$ cells.

Neighbor lists: The idea here is similar to the above, but better refined. Instead of creating a "list" for a cluster of particles, we create one for each particle. Thus, the idea goes like: For particle i, make a list of all particles in the system that can interact with i. Normally, particles do not move far within a time step, in order to avoid creating the lists every time step, we record particles in the sphere of radius $r_c + \epsilon$ instead of just c. This ϵ is an adjustable parameter (depending on physics and computing resources).

Of course, to search particles in the entire domain to create the lists is just too costly. The best method is to use particle-in-cell to create approximate lists and then to create the neighbors lists.

These two methods are very useful; we call them screening method.

I. Particle decomposition: Particle decomposition method distributes particles, as in long-ranged interaction (with no consideration for the particles' positions), uniformly to all processors. This method corresponds to force matrix row partition.

Step I : Screening.
Step II : Communicate boarder mesh particles to relevant processors.
Step III: Update particle positions in all processors.
Step IV: Scatter the new particles (same particles but new positions).

Remarks

- Load balancing requires the system being of uniform density (although each processor has the same amount of particles, but each particle may not have the same number of neighbors). A simple way to relax this problem is to re-distribute the particles to processors randomly.
- Global communication is needed for screening and scattering particles after updates.
- Simple to implement.

II. Force decomposition: Force decomposition corresponds to matrix sub-blocking. Each processor is assigned a sub-block $F_{\alpha\beta}$ of the force matrix to evaluate. For a system of N particles by p processors, each the force sub-matrix is of size $\frac{N}{\sqrt{p}} \times \frac{N}{\sqrt{p}}$.

Algorithm

(1) Get ready to receive the message data from the source processor by calling `irecv`.
(2) Send the message data to the destination processor by calling `csend`.
(3) Wait for message.

Force computing

(1) Screening.
(2) Compute $F_{\alpha\beta}$ and sum over all β to find the partial force on particles α by particles β.
(3) Processor $P_{\alpha\beta}$ collects the partial forces from row α processors to compute the total force on particles α.
(4) Update the N/p particles within a group α.
(5) Broadcast new positions to all other processors.
(6) Repeat Steps 1–4 to the next time step.

Remarks

- Load imbalance is again a serious problem because the matrix density is obviously non-uniform.
- No geometric information is needed for decomposition.

Space decomposition: Space decomposition is a tricker method. Basically, we slice the computational domain into p sub-domains. These sub-domain boarder lines are made to be co-linear with the cell lines. Typically each cell contains about 10 particles (of course, it is physics-dependent.) and the number of cells each sub-domain contains depends on N and p.

This method is similar to solving hyperbolic PDE on parallel processors, i.e. building buffer zones.

(1) Partition particles into participating processors according to particles' positions.

(2) Construct buffer zone for communication. Then communicate the buffer zones to relevant processors to build the extended sub-domains on each processor.

(3) Compute the forces for particles in each physical box (in parallel) and update their positions. There is no need to update the positions of the particles in the buffer zones.

(4) Check the new positions of the particles which lie in the buffer zone at t_{old}.

(5) Repeat Steps 2–4.

Remarks

- Load imbalance is also a serious problem. To solve this problem, we need to decompose the computational domain into non-uniform sub-domains (the decomposition is also dynamic). This way, one can minimize the load imbalance effect.
- If a reasonable load balance is achieved, this is a very fast method.

10.2. Schrödinger's Equations and Quantum Mechanics

The basic equation by Ervin Schrödinger (1887–1961) in quantum mechanics which describes the evolution of atomic-scale motions is given by:

$$\frac{h^2}{8\pi^2 m}\nabla^2\Psi(r,t) + V\Psi(r,t) = -\frac{h}{2\pi i}\frac{\partial\Psi(r,t)}{\partial t} \tag{10.12}$$

10.3. Partition Function, DFT and Material Science

The partition function is a central construct in statistics and statistical mechanics. It also serves as a bridge between thermodynamics and quantum mechanics because it is formulated as a sum over the states of a macroscopic system at a given temperature. It is commonly used in condensed-matter physics and in theoretical studies of high Tc superconductivity.

$$Z = \sum_{j} g_j e^{-\frac{E_j}{KT}} \tag{10.13}$$

10.3.1. *Materials research*

Topics include material property prediction, modeling of new materials, and superconductivity.

Material property: High-performance computing has provided invaluable assistance in improving our understanding of the atomic nature of materials. A selected list of such materials includes semiconductors, such as silicon and gallium arsenide, and superconductors such as the high-copper oxide ceramics that have been shown recently to conduct electricity at extremely high temperatures.

Superconductivity: The discovery of high-temperature superconductivity in 1986 has provided the potential of spectacular energy-efficient power transmission technologies, ultra-sensitive instrumentation, and devices using phenomena unique to superconductivity. The materials supporting high temperature-superconductivity are difficult to form, stabilize, and use, and the basic properties of the superconductor must be elucidated through a vigorous fundamental research program.

Semiconductor devices design: As intrinsically faster materials such as gallium arsenide are used, a fundamental understanding is required of how they operate and how to change their characteristics. Essential understanding of overlay formation, trapped structural defects, and the effect of lattice mismatch on properties are needed. Currently, it is possible to simulate electronic properties for simple regular systems; however, materials with defects and mixed atomic constituents are beyond present capabilities. Simulating systems with 100 million particles is possible on the largest parallel computer — 2000-node Intel Paragon.

Nuclear fusion: Development of controlled nuclear fusion requires understanding the behavior of fully ionized gasses (plasma) at very high temperatures under the influence of strong magnetic fields in complex 3D geometries.

10.4. Maxwell's Equations and Electrical Engineering

The equations by James Clerk Maxwell (1831–1879) describe the relationship between electric and magnetic fields at any point in space as a function of charge and electric current densities at such a point. The wave equations for the propagation of light can be derived from these equations, and they are the basic equations for the classical electrodynamics. They are used in the studies of most electromagnetic phenomena including plasma physics as well as the earth's magnetic fields and its interactions with other fields in the cosmos.

$$\nabla \times E = -\frac{\partial B}{\partial t}$$
$$\nabla \cdot D = \rho$$
$$\nabla \times H = \frac{\partial D}{\partial t} + J \qquad (10.14)$$
$$\nabla \cdot B = 0$$

10.4.1. *Helmholtz equation*

This equation, discovered by Herman von Helmholtz (1821–1894), is used in acoustics and electromagnetism. It is also used in the study of vibrating membranes.

$$-\Delta u + \lambda = f \qquad (10.15)$$

10.4.2. *Electrical engineering*

Topics include electromagnetic scattering, wireless communication, and antenna design.

10.5. Diffusion Equation and Mechanical Engineering

This is the equation that describes the distribution of a certain field variable such as temperature as a function of space and time. It is used to explain physical, chemical, reaction-diffusion systems, as well as some biological phenomena.

$$\Delta u = \frac{\partial u}{\partial t} \qquad (10.16)$$

Topics include structural analysis, combustion simulation, and vehicle simulation.

Vehicle dynamics: Analysis of the aero-elastic behavior of vehicles, and the stability and ride analysis of vehicles are critical assessments of land and air vehicle performance and life cycle.

Combustion systems: Attaining significant improvements in combustion efficiencies require understanding the interplay between the flows of the various substances involved and the quantum chemistry that causes those substances to react. In some complicated cases, the quantum chemistry is beyond the reach of current supercomputers.

10.6. Navier-Stokes Equation and CFD

The Navier-Stokes equation, developed by Claude Louis Marie Navier (1785–1836) and Sir George Gabriel Stokes (1819–1903), is the primary equation of computational fluid dynamics. It relates the pressure and external forces acting on a fluid to the response of the fluid flow. Forms of this equation are used in computations for aircraft and ship design, weather prediction, and climate modeling.

$$\frac{\partial u}{\partial t} + (u \cdot \nabla)u = -\frac{1}{\rho}\nabla p + \gamma \nabla^2 u + \frac{1}{\rho}F \qquad (10.17)$$

Aircraft design, air-breathing propulsion, advanced sensors are some examples.

10.7. Other Applications

10.7.1. *Astronomy*

Data volumes generated by very large array or very long baseline array radio telescopes currently overwhelm the available computational resources. Greater computational power will significantly enhance their usefulness in exploring important problems in radio astronomy.

10.7.2. *Biological engineering*

Current areas of research include simulation of genetic compounds, neural networks, structural biology, conformation, drug design, protein folding, and the human genome.

Structural biology: The function of biologically important molecules can be simulated by computationally intensive Monte Carlo methods in combination with crystallographic data derived from nuclear magnetic resonance measurements. Molecular dynamics methods are required for the time dependent behavior of such macromolecules. The determination, visualization, and analysis of these 3D structures are essential to the understanding of the mechanisms of enzymic catalysis, recognition of nucleic acids by proteins, antibody and antigen binding, and many other dynamic events central to cell biology.

Design of drugs: Predictions of the folded conformation of proteins and of RNA molecules by computer simulation are rapidly becoming accepted as useful, and sometimes as a primary tool in understanding the properties required in drug design.

Human genome: Comparison of normal and pathological molecular sequences is currently our most revealing computational method for understanding genomes and the molecular basis for disease. To benefit from the entire sequence of a single human will require capabilities for more than three billion sub-genomic units, as contrasted with the 10 to 200,000 units of typical viruses.

10.7.3. *Chemical engineering*

Topics include polymer simulation, reaction rate prediction, and chemical vapor deposition for thin-films.

10.7.4. *Geosciences*

Topics include oil and seismic exploration, enhanced oil and gas recovery.

Enhanced oil and gas recovery: This challenge has two parts. First, one needs to locate the estimated billions of barrels of oil reserves on the earth and then to devise economic ways of extracting as much of it as possible. Thus, improved seismic analysis techniques in addition to improved understanding of fluid flow through geological structures are required.

10.7.5. *Meteorology*

Topics include prediction of Weather, Climate, Typhoon, and Global change. The aim here is to understand the coupled atmosphere-ocean, biosphere system in enough detail to be able to make long-range predictions about its behavior. Applications include understanding dynamics in the atmosphere, ozone depletion, climatological perturbations owing to man-made releases of chemicals or energy into one of the component systems, and detailed predictions of conditions in support of military missions.

10.7.6. *Oceanography*

Ocean sciences: The objective is to develop a global ocean predictive model incorporating temperature, chemical composition, circulation, and coupling to the atmosphere and other oceanographic features. This ocean model will couple to models of the atmosphere in the effort on global weather and have specific implications for physical oceanography as well.

APPENDIX A

MPI

A.1. An MPI Primer

A.1.1. *What is MPI?*

MPI is the standard for multi-computer and cluster message passing introduced by the Message-Passing Interface Forum in April 1994. The goal of MPI is to develop a widely used standard for writing message-passing programs.

MPI plays an intermediary between parallel programming language and specific running environment like other portability helpers. MPI is more widely accepted as a portability standard.

MPI was implemented by some vendors on different platforms. Suitable implementations of MPI can be found for most distributed-memory systems.

To the programmer, MPI appears in the form of libraries for FORTRAN or C family languages. Message passing is realized by an MPI routine (or function) call.

A.1.2. *Historical perspective*

Many developers contributed to the MPI library, as shown in Fig. A.1.

A.1.3. *Major MPI issues*

Process creation and management: Discusses the extension of MPI to remove the static process model in MPI. It defines routines that allow for creation of processes.

Cooperative and One-Sided Communications: One worker performs transfer of data (the opposite of cooperative). It defines communication routines that can be completed by a single process. These include shared-memory operations (put/get) and remote accumulate operations.

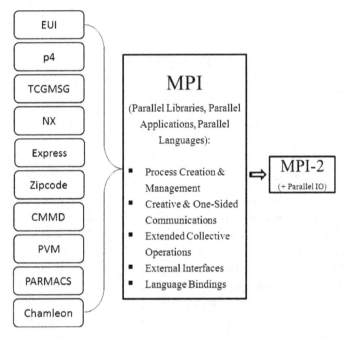

Fig. A.1.   Evolution and key components of MPI.

Extended collective operations: This extends the semantics of MPI-1 collective operations to include intercommunicators. It also adds more convenient methods of constructing intercommunicators and two new collective operations.

External interfaces: This defines routines designed to allow developers to layer on top of MPI. This includes generalized requests, routines that decode MPI opaque objects, and threads.

I/O: MPI-2 support for parallel I/O.

Language Bindings: C++ binding and FORTRAN-90 issues.

A.1.4. *Concepts and terminology for message passing*

Distributed-Memory: Every processor has its own local memory which can be accessed directly only by its own CPU. Transfer of data from one processor to another is performed over a network. This differs from shared-memory systems which permit multiple processors to directly access the same memory resource via a memory bus Fig. A.2.

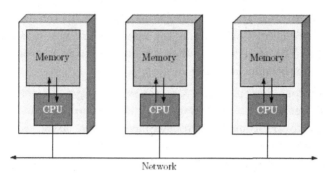

Fig. A.2. This figure illustrates the structure of a distributed-memory system.

Message Passing: This is the method by which data from one processor's memory is copied to the memory of another processor. In distributed-memory systems, data is generally sent as packets of information over a network from one processor to another. A message may consist of one or more packets, and usually includes routing and/or other control information.

Process: A process is a set of executable instructions (a program) which runs on a processor. One or more processes may execute on a processor. In a message passing system, all processes communicate with each other by sending messages, even if they are running on the same processor. For reasons of efficiency, however, message passing systems generally associate only one process per processor.

Message passing library: This usually refers to a collection of routines which are imbedded in application code to accomplish send, receive and other message passing operations.

Send/Receive: Message passing involves the transfer of data from one process (send) to another process (receive), requires the cooperation of both the sending and receiving process. Send operations usually require the sending process to specify the data's location, size, type and the destination. Receive operations should match a corresponding send operation.

Synchronous/Asynchronous: Synchronous send operations will complete only after the acknowledgement that the message was safely received by the receiving process. Asynchronous send operations may "complete" even though the receiving process has not actually received the message.

Application buffer: It is the address space that holds the data which is to be sent or received. For example, suppose your program uses a variable called inmsg. The application buffer for inmsg is the program memory location where the value of inmsg resides.

System buffer: System space for storing messages. Depending upon the type of send/receive operation, data in the application buffer may be required to be copied to/from system buffer space. This allows communication to be asynchronous.

Blocking communication: A communication routine is blocking if the completion of the call is dependent on certain "events". For sends, the data must be successfully sent or safely copied to system buffer space so that the application buffer that contained the data is available for reuse. For receives, the data must be safely stored in the receive buffer so that it is ready for use.

Non-blocking communication: A communication routine is non-blocking if the call returns without waiting for any communications events to complete (such as copying of message from user memory to system memory or arrival of message). It is not safe to modify or use the application buffer after completion of a non-blocking send. It is the programmer's responsibility to insure that the application buffer is free for reuse. Non-blocking communications are primarily used to overlap computation with communication to provide gains in performance.

Communicators and Groups: MPI uses objects called communicators and groups to define which collection of processes may communicate with each other. Most MPI routines require us to specify a communicator as an argument. Communicators and groups will be covered in more detail later. For now, simply use MPI COMM WORLD whenever a communicator is required — it is the predefined communicator which includes all of our MPI processes.

Rank: Within a communicator, every process has its own unique integer identifier assigned by the system when the process initializes. A rank is sometimes also called a "process ID." Ranks are contiguous and begin at zero. In addition, it is used by the programmer to specify the source and destination of messages, and is often used conditionally by the application to control program execution. For example,

$$\begin{cases} \text{rank} = 0, & \text{do this} \\ \text{rank} = 1, & \text{do that} \end{cases}$$

Message attributes

(1) The envelope.
(2) Rank of destination.
(3) Message tag.

  (a) ID for a particular message to be matched by both sender and receiver.
  (b) It is like sending multiple gifts to your friend; you need to identify them.
  (c) MPI TAG UB $\leq$ 32767.
  (d) Similar in functionality to "comm" to group messages.
  (e) "comm" is safer than "tag", but "tag" is more convenient.

(4) Communicator.
(5) The Data.
(6) Initial address of send buffer.
(7) Number of entries to send.
(8) Datatype of each entry.

  (a) `MPI_INTEGER`
  (b) `MPI_REAL`
  (c) `MPI_DOUBLE PRECISION`
  (d) `MPI_COMPLEX`
  (e) `MPI_LOGICAL`
  (f) `MPI_BYTE`
  (g) `MPI_INT`
  (h) `MPI_CHAR`
  (i) `MPI_FLOAT`
  (j) `MPI_DOUBLE`

A.1.5. *Using the MPI programming language*

All the MPI implementations nowadays support FORTRAN and C. Implementations of MPI-2 also support C++ and FORTRAN 90. Any one of these languages can be used with the MPI library to produce parallel programs. However, there are a few differences in the MPI call between the FORTRAN and C families.

Header File

```
      C                   FORTRAN
#include ''mpi.h''   Include 'mpif.h'
```

MPI call format

Language	C	FORTRAN
Format	rc = MPI_Xxxxx	CALL MPI_XXXXX
	(parameter,...)	(parameter,...,ierr)
Example	rc = MPI_Bsend	CALL MPI_BSEND
	(&buf, count, type,	(buf, count, type, dest,
	dest, tag, comm)	tag, comm, ierr)
Error	Returned as rc	Returned as ierr parameter
Code	MPI_SUCCESS if successful	MPI_SUCCESS if successful

Program structure: The general structures are the same in both language families. These four components must be included in the program:

(1) Include the MPI header.
(2) Initialize the MPI environment.
(3) Normal language sentences and MPI calls.
(4) Terminate the MPI environment.

A.1.6. *Environment management routines*

Several of the more commonly used MPI environment management routines are described below:

MPI_Init

MPI_Init initializes the MPI execution environment. This function must be called in every MPI program, before any other MPI functions, and must be called only once in an MPI program. For C programs, MPI_Init may be used to pass the command line arguments to all processes, although this is not required by the standard and is implementation dependent.

```
MPI_Init(*argc,*argv)
MPI_INIT(ierr)
```

MPI_Comm_size

This determines the number of processes in the group associated with a communicator. It is generally used within the communicator MPI_COMM WORLD to determine the number of processes being used by our application.

```
MPI_Comm_size(comm,*size)
MPI_COMM_SIZE(comm,size,ierr)
```

MPI_Comm_rank

MPI_Comm rank determines the rank of the calling process within the communicator. Initially, each process will be assigned a unique integer rank between 0 and P−1, within the communicator MPI_COMM WORLD. This rank is often referred to as a task ID. If a process becomes associated with other communicators, it will have a unique rank within each of these as well.

MPI_Comm_rank(comm,*rank)
MPI_COMM_RANK(comm,rank,ierr)

MPI_Abort

This function terminates all MPI processes associated with a communicator. In most MPI implementations, it terminates all processes regardless of the communicator specified.

MPI_Abort(comm,errorcode)
MPI_ABORT(comm,errorcode,ierr)

MPI_Get_processor_name

This gets the name of the processor on which the command is executed. It also returns the length of the name. The buffer for name must be at least MPI_MAX_PROCESSOR_NAME characters in size. What is returned into name is implementation dependent — it may not be the same as the output of the hostname or host shell commands.

MPI_Get_processor_name(*name,*resultlength)
MPI_GET_PROCESSOR_NAME(name, resultlength, ierr)

MPI_Initialized

MPI_Initialized indicates whether MPI_Init has been called and returns a flag as either true (1) or false (0). MPI requires that MPI_Init be called once and only once by each process. This may pose a problem for modules that want to use MPI and are prepared to call MPI_Init if necessary. MPI Initialized solves this problem.

MPI_Initialized(*flag)
MPI_INITIALIZED(flag,ierr)

MPI_Wtime

This returns an elapsed wall clock time in seconds (double precision) on the calling processor.

`MPI_Wtime()`
`MPI_WTIME()`

MPI_Wtick

This returns the resolution in seconds (double precision) of `MPI_Wtime`.

`MPI_Wtick()`
`MPI_WTICK()`

MPI_Finalize

Terminates the MPI execution environment. This function should be the last MPI routine called in every MPI program, no other MPI routines may be called after it.

`MPI_Finalize()`
`MPI_FINALIZE(ierr)`

Examples: Figures A.3 and A.4 provide some simple examples of environment management routine calls.

```
#include"mpi.h"
#include<stdio.h>

int main(int argc, char *argv[]){
int numtasks, rank, rc;

rc = MPI_Init(&argc, &argv);
if (rc != 0){
    printf("Error starting MPI program. Terminating.\n");
    MPI_Abort(MPI_COMM_WORLD, rc);
    }

MPI_Comm_size(MPI_COMM_WORLD, &numtasks);
MPI_Comm_rank(MPI_COMM_WORLD, &rank);
printf("Number of tasks= %d My rank= %d\n", numtasks, rank);

/******  do some work ******/

MPI_Finalize();
return 0;
}
```

Fig. A.3.   A simple example of environment management in C.

```
program simple
include 'mpif.h'
integer numtasks, rank, ierr, rc
call MPI_INIR(ierr)
if (ierr .ne. 0) then
    print *, 'Error starting MPI program. Terminating.'
    call MPI_ABORT(MPI_COMM_WORLD, rc, ierr)
end if
call MPI_COMM_RANK(MPI_COMM_WORLD, rank, ierr)
call MPI_COMM_SIZE(MPI_COMM_WORLD, numtasks, ierr)
print *, 'Number of tasks=', numtasks, ' My rank=' ,rank

C ****** do some work ******

call MPI_FINALIZE(ierr)
end
```

Fig. A.4. A simple example of environment management in FORTRAN.

A.1.7. *Point to point communication routines*

Blocking message passing routines

The more commonly used MPI blocking message passing routines are described on p. 152.

MPI_Send

As a basic blocking send operation, this routine returns only after the application buffer in the sending task is free for reuse. Note that this routine may be implemented differently on different systems. The MPI standard permits the use of a system buffer but does not require it. Some implementations may actually use a synchronous send (discussed below) to implement the basic blocking send.

MPI_Send(*buf,count,datatype,dest,tag,comm)

MPI_SEND(buf, count, data type, dest, tag, comm, ierr)

Buffer: This is the program (application) address space which references the data that is to be sent or received. In most cases, this is simply the variable name that is being sent/received. For C programs, this argument is passed by reference and usually must be prepended with an ampersand (&var1).

Data Count: Indicates the number of data elements of a particular type to be sent.

Data Type: For reasons of portability, MPI predefines its data types. Programmers may also create their own data types (derived types). Note

Table A.1.   MPI data types.

MPI C data types		MPI FORTRAN data types	
MPI_CHAR	signed char	MPI_CHARACTER	character(1)
MPI_SHORT	singed chort int		
MPI_INT	signed int	MPI_INTEGER	integer
MPI_LONG	signed long int		
MPI_UNSIGNED_CHAR	unsigned char		
MPI_UNSIGNED_SHORT	unsigned short int		
MPI_UNSIGNED	unsigned int		
MPI_UNSIGNED_LONG	unsigned int		
MPI_FLOAT	float	MPI_REAL	real
MPI_DOUBLE	double	MPI_DOUBLE_PRECISION	double precision
MPI_LONG_DOUBLE	long double		
		MPI_COMPLEX	complex
		MPI_LOGICAL	logical
MPI_BYTE	8 binary digits	MPI_BYTE	8 binary digits
MPI_PACKED	data packed or unpacked with MPI_Pack()/ MPI_Unpack	MPI_PACKED	data packed or unpacked with MPI_Pack()/ MPI_Unpack

that the MPI types `MPI_BYTE` and `MPI_PACKED` do not correspond to standard C or FORTRAN types. Table A.1 lists the MPI data types for both C and FORTRAN.

Destination: This is an argument to send routines which indicates the process where a message should be delivered. It is specified as the rank of the receiving process.

Source: For `MPI_Recv`, there is an argument source corresponding to destination in `MPI_Send`. This is an argument to receive routines which indicates the originating process of the message. Specified as the rank of the sending process, it may be set to the wild card `MPI_ANY_SOURCE` to receive a message from any task.

Tag: An arbitrary, non-negative integer assigned by the programmer to uniquely identify a message. Send and receive operations should match message tags. For a receive operation, the wild card ANY TAG can be used to receive any message regardless of its tag. The MPI standard guarantees that integers from the range [0, 32767] can be used as tags, but most implementations allow a much larger range.

Communicator: Indicates the communication context, or set of processes for which the source or destination fields are valid. Unless the programmer is explicitly creating new communicators, the predefined communicator MPI_COMM_WORLD is usually used.

Status: For a receive operation, this indicates the source of the message and the tag of the message. In C, this argument is a pointer to a predefined structure MPI_Status (ex. stat.MPI_SOURCE stat.MPI_TAG). In FORTRAN, it is an integer array of size MPI_STATUS_SIZE (ex. stat(MPI_SOURCE) stat(MPI_TAG)). Additionally, the actual number of bytes received are obtainable from status via the MPI_Get_count_routine.

Request: This is used by non-blocking send and receive operations. Since non-blocking operations may return before the requested system buffer space is obtained, the system issues a unique "request number." The programmer uses this system assigned "handle" later (in a WAIT type routine) to determine completion of the non-blocking operation. In C, this argument is a pointer to a predefined structure MPI_Request. In FORTRAN, it is an integer.

MPI_Recv

Receive a message and block until the requested data is available in the application buffer in the receiving task.

MPI_Recv(*buf,count,datatype,source,tag,comm,*status)
MPI_RECV(buf,count,datatype,source,tag,comm,status,ierr)

MPI_Ssend

Synchronous blocking send: Send a message and block until the application buffer in the sending task is free for reuse and the destination process has started to receive the message.

MPI_Ssend(*buf,count,datatype,dest,tag,comm,ierr)
MPI_SSEND(buf,count,datatype,dest,tag,comm,ierr)

MPI_Bsend

Buffered blocking send: Permits the programmer to allocate the required amount of buffer space into which data can be copied until it is delivered. It insulates against the problems associated with insufficient system buffer space. Routine returns after the data has been copied from application buffer space to the allocated send buffer. This must be used with the MPI_Buffer_attach routine.

MPI_Bsend(*buf,count,datatype,dest,tag,comm)
MPI_BSEND(buf,count,datatype,dest,tag,comm,ierr)

MPI_Buffer_attach, MPI_Buffer_detach

Used by the programmer to allocate/deallocate message buffer space to be used by the MPI_Bsend routine. The size argument is specified in actual data bytes — not a count of data elements. Only one buffer can be attached to a process at a time. Note that the IBM implementation uses MPI_BSEND_OVERHEAD bytes of the allocated buffer for overhead.

```
MPI_Buffer_attach(*buffer,size)
MPI_Buffer_detach(*buffer,size)
MPI_BUFFER_ATTACH(buffer,size,ierr)
MPI_BUFFER_DETACH(buffer,size,ierr)
```

MPI_Rsend

A blocking ready send should only be used if the programmer is certain that the matching receive has already been posted.

```
MPI_Rsend(*buf,count,datatype,dest,tag,comm)
MPI_RSEND(buf,count,datatype,dest,tag,comm,ierr)
```

MPI_Sendrecv

Send a message and post a receive before blocking. This will block until the sending application buffer is free for reuse and until the receiving application buffer contains the received message.

```
MPI_Sendrecv(*sendbuf,sendcount,sendtype,dest,sendtag,
*recv_buf,recvcount,recvtype,source,recvtag,comm,*status)
MPI_SENDRECV(sendbuf,sendcount,sendtype,dest,sendtag,
recvbuf,recvcount,recvtype,source,recvtag,comm,status,ierr)
```

MPI_Probe

Performs a blocking test for a message. The "wild cards" MPI_ANY_SOURCE and MPI_ANY_TAG may be used to test for a message from any source or with any tag. For the C routine, the actual source and tag will be returned in the status structure as status.MPI_SOURCE and status.MPI_TAG. For the FORTRAN routine, they will be returned in the integer array status(MPI_SOURCE) and status(MPI_TAG).

```
MPI_Probe(source,tag,comm,*status)
MPI_PROBE(source,tag,comm,status,ierr)
```

Examples: Figures. A.5 and A.6 provide some simple examples of blocking message passing routine calls.

```
#include "mpi.h"
#include <stdio.h>

int main(int argc, char *argv[]){
int numtasks, rank, dest, source, rc, tag=1;
char inmsg, outmsg='x';
MPI_Status Stat;

MPI_Init(&argc, &argv);
MPI_Comm_size(MPI_COMM_WORLD, &numtasks);
MPI_Comm_rank(MPI_COOM_World, &rank);

if (rank == 0){
    dest = 1;
    source = 1;
    rc = MPI_Send(&outmsg, 1, MPI_CHAR, dest, tag,
        MPI_COMM_WORLD);
    rc = MPI_Recv(&inmsg, 1, MPI_CHAR, source, tag,
        MPI_COMM_WORLD, &Stat);
    }

else if (rank == 1){
    dest = 0;
    source = 1;
    rc = MPI_Recv(&inmsg, 1, MPI_CHAR, source, tag,
        MPI_COMM_WORLD, &Stat);
    rc = MPI_Send(&outmsg, 1, MPI_CHAR, dest, tag,
        MPI_COMM_WORLD);
    }

MPI_Finalize();
return 0;
}
```

Fig. A.5. A simple example of blocking message passing in C. Task 0 pings task 1 and awaits a return ping.

Non-blocking message passing routines

The more commonly used MPI non-blocking message passing routines are described below.

MPI_Isend

This identifies an area in memory to serve as a send buffer. Processing continues immediately without waiting for the message to be copied out from the application buffer. A communication request handle is returned for handling the pending message status. The program should not modify the

```
      program ping
      include 'mpif.h'
      integer numtasks, rank, dest, source, tag, ierr
      integer stat(MPI_STATUS_SIZE)
      character inmsg, outmsg
      tag = 1
      call MPI_INIT(ierr)
      call MPI_COMM_RANK(MPI_COMM_WORLD, rank, ierr)
      call MPI_COMM_SIZE(MPI_COMM_WORLD, numtasks, ierr)
      if (rank .eq. 0) then
          dest = 1
          source = 1
          outmsg = 'x'
          call MPI_SEND(outmsg, 1, MPI_CHARACTER, dest, tag,
     &          MPI_COMM_WORLD, ierr)
          call MPI_RECV(inmsg, 1, MPI_CHARACTER, source, tag,
     &          MPI_COMM_WORLD, stat, ierr)
      else if (rank .eq. 1) then
          dest = 1
          source = 1
          outmsg = 'x'
          call MPI_RECV(inmsg, 1, MPI_CHARACTER, source, tag,
     &          MPI_COMM_WORLD, stat, ierr)
          call MPI_SEND(outmsg, 1, MPI_CHARACTER, dest, tag,
     &          MPI_COMM_WORLD, ierr)
      endif
      call MPI_FINALIZE(ier)
      end
```

Fig. A.6.   A simple example of blocking message passing in FORTRAN.

application buffer until subsequent calls to MPI_Wait or MPI_Test indicates that the non-blocking send has completed.

MPI_Isend(*buf,count,datatype,dest,tag,comm,*request)
MPI_ISEND(buf,count,datatype,dest,tag,comm,request,ierr)

MPI_Irecv

This identifies an area in memory to serve as a receive buffer. Processing continues immediately without actually waiting for the message to be received and copied into the application buffer. A communication request handle is returned for handling the pending message status. The program must use calls to MPI_Wait or MPI_Test to determine when the non-blocking receive operation completes and the requested message is vailable in the application buffer.

MPI_Irecv(*buf,count,datatype,source,tag,comm,*request)
MPI_IRECV(buf,count,datatype,source,tag,comm,request,ierr)

MPI_Issend

Non-blocking synchronous send. Similar to MPI_Isend(), except MPI_Wait() or MPI_Test() indicates when the destination process has received the message.

```
MPI_Issend(*buf,count,datatype,dest,tag,comm,*request)
MPI_ISSEND(buf,count,datatype,dest,tag,comm,request,ierr)
```

MPI_Ibsend

A non-blocking buffered send that is similar to MPI_Bsend() except MPI_Wait() or MPI_Test() indicates when the destination process has received the message. Must be used with the MPI_Buffer attach routine.

```
MPI_Ibsend(*buf,count,datatype,dest,tag,comm,*request)
MPI_IBSEND(buf,count,datatype,dest,tag,comm,request,ierr)
```

MPI_Irsend

Non-blocking ready send is similar to MPI_Rsend() except MPI_Wait() or MPI_Test() indicates when the destination process has received the message. This function should only be used if the programmer is certain that the matching receive has already been posted.

```
MPI_Irsend(*buf,count,datatype,dest,tag,comm,*request)
MPI_IRSEND(buf,count,datatype,dest,tag,comm,request,ierr)
```

MPI_Test, MPI_Testany, MPI_Testall, MPI_Testsome

MPI_Test checks the status of a specified non-blocking send or receive operation. The "flag" parameter is returned logical true (1) if the operation has completed, and logical false (0) if not. For multiple non-blocking operations, the programmer can specify any, all or some completions.

```
MPI_Test(*request,*flag,*status)
MPI_Testany(count,*array_of_requests,*index,*flag,*status)
MPI_Testall(count,*array_of_requests,*flag,
    *array_of_statuses)
MPI_Testsome(incount,*array_of_requests,*outcount,
    *array_of_offsets, *array_of_statuses)
MPI_TEST(request,flag,status,ierr)
MPI_TESTANY(count,array_of_requests,index,flag,status,ierr)
MPI_TESTALL(count,array_of_requests,flag,array_of_statuses,
    ierr)
```

```
MPI_TESTSOME(incount,array_of_requests,outcount,
    array_of_offsets, array_of_statuses,ierr)
```

MPI_Wait, MPI_Waitany, MPI_Waitall, MPI_Waitsome

MPI_Wait blocks until a specified non-blocking send or receive operation has completed. For multiple non-blocking operations, the programmer can specify any, all or some completions.

```
MPI_Wait(*request,*status)
MPI_Waitany(count,*array_of_requests,*index,*status)
MPI_Waitall(count,*array_of_requests,*array_of_statuses)
MPI_Waitsome(incount,*array_of_requests,*outcount,
    *array_of_offsets,*array_of_statuses)
MPI_WAIT(request,status,ierr)
MPI_WAITANY(count,array_of_requests,index,status,ierr)
tt MPI_WAITALL(count,array_of_requests,array_of_statuses,ierr)
MPI_WAITSOME(incount,array_of_requests,outcount,
    array_of_offsets, array_of_statuses,ierr)
```

MPI_Iprobe

Performs a non-blocking test for a message. The "wildcards" MPI_ANY_SOURCE and MPI_ANY_TAG may be used to test for a message from any source or with any tag. The integer "flag" parameter is returned logical **true** (1) if a message has arrived, and logical **false** (0) if not. For the C routine, the actual source and tag will be returned in the status structure as status.MPI_SOURCE and status.MPI_TAG. For the FORTRAN routine, they will be returned in the integer array status(MPI_SOURCE) and status(MPI_TAG).

```
MPI_Iprobe(source,tag,comm,*flag,*status)
MPI_IPROBE(source,tag,comm,flag,status,ierr)
```

Examples: Figures A.7 and A.8 provide some simple examples of blocking message passing routine calls.

A.1.8. *Collective communication routines*

Collective communication involves all processes in the scope of the communicator. All processes are by default, members in the communicator MPI_COMM_WORLD.

There are three types of collective operations:

(1) **Synchronization:** Processes wait until all members of the group have reached the synchronization point.

```
#include "mpi.h"
#include <stdio.h>

int main(int argc, char *argv[]){
int numtasks, rank, next, prev, buf[2], tag1=1, tag2=2;
MPI_Request reqs[4];
MPI_Status stats[4];

MPI_Init(&argc, &argv);
MPI_Comm_size(MPI_COMM_WORLD, &numtasks);
MPI_Comm_rank(MPI_COMM_WORLD, &rank);

prev = rank-1;
next = rank+1;
if (rank == 0) prev = numtasks - 1;
if (rank == (numtasks - 1)) next = 0;

MPI_Irecv(&buf[0], 1, MPI_INT, prev, tag1, MPI_COMM_WORLD,
    &reqs[0]);
MPI_Irecv(&buf[1], 1, MPI_INT, next, tag2, MPI_COMM_WORLD,
    &reqs[1]);

MPI_Irecv(&rank, 1, MPI_INT, prev, tag2, MPI_COMM_WORLD,
    &reqs[2]);
MPI_Irecv(&rank, 1, MPI_INT, prev, tag1, MPI_COMM_WORLD,
    &reqs[3]);

MPI_Waitall(4, reqs, stats);

MPI_Finalize();
return 0;
}
```

Fig. A.7. A simple example of non-blocking message passing in C, this code represents a nearest neighbor exchange in a ring topology.

(2) **Data movement:** Broadcast, scatter/gather, all to all.

(3) **Collective computation** (reductions): One member of the group collects data from the other members and performs an operation (min, max, add, multiply, etc.) on that data.

Collective operations are blocking. Collective communication routines do not take message tag arguments. Collective operations within subsets of processes are accomplished by first partitioning the subsets into a new groups and then attaching the new groups to new communicators (discussed later). Finally, work with MPI defined datatypes — not with derived types.

MPI_Barrier

```
      program ringtopo
      include 'mpif.h'
      integer numtasks, rank, next, prev, buf(2), tag1, tag2
      integer ierr, stats(MPI_STATUS_SIZE,4), reqs(4)
      tag1 = 1
      tag2 = 2
      call MPI_INIT(ierr)
      call MPI_COMM_RANK(MPI_COMM_WORLD, rank, ierr)
      call MPI_COMM_SIZE(MPI_COMM_WORLD, numtasks, ierr)
      prev = rank - 1
      next = rank + 1
      if (rank .eq. 0) then
         prev = numtasks - 1
      endif
      if (rank .eq. numtasks - 1) then
         next = 0
      endif

      call MPI_IRECV(buf(1), 1, MPI_INTEGER, prev, tag1,
     &       MPI_COMM_WORLD, reqs(1), ierr)
      call MPI_IRECV(buf(2), 1, MPI_INTEGER, prev, tag2,
     &       MPI_COMM_WORLD, reqs(2), ierr)

      call MPI_ISEND(buf(1), 1, MPI_INTEGER, prev, tag1,
     &       MPI_COMM_WORLD, reqs(1), ierr)
      call MPI_ISEND(buf(2), 1, MPI_INTEGER, prev, tag2,
     &       MPI_COMM_WORLD, reqs(2), ierr)
      call MPI_WAITALL(4, reqs, stats, ierr)
      call MPI_FINALIZE(ierr)
      end
```

Fig. A.8.   A simple example of non-blocking message passing in FORTRAN.

Creates a barrier synchronization in a group. Each task, when reaching the MPI_Barrier call, blocks until all tasks in the group reach the same MPI_Barrier call.

MPI_Barrier(comm)
MPI_BARRIER(comm,ierr)

MPI_Bcast

Broadcasts (sends) a message from the process with rank "root" to all other processes in the group.

MPI_Bcast(*buffer,count,datatype,root,comm)
MPI_BCAST(buffer,count,datatype,root,comm,ierr)

MPI_Scatter

Distributes distinct messages from a single source task to each task in the group.

```
MPI_Scatter(*sendbuf,sendcnt,sendtype,*recvbuf,ecvcnt,
    recvtype,root,comm)
MPI_SCATTER(sendbuf,sendcnt,sendtype,recvbuf,recvcnt,
    recvtype,root,comm,ierr)
```

MPI_Gather

Gathers distinct messages from each task in the group to a single destination task. This routine is the reverse operation of MPI_Scatter.

```
MPI_Gather(*sendbuf,sendcnt,sendtype,recvbuf,recvcount,
    recvtype,root,comm)
MPI_GATHER(sendbuf,sendcnt,sendtype,recvbuf,recvcount,
    recvtype,root,comm,ierr)
```

MPI_Allgather

Concatenation of data to all tasks in a group. Each task in the group, in effect, performs a one-to-all broadcasting operation within the group.

```
MPI_Allgather(*sendbuf,sendcount,sendtype,recvbuf,
    recvcount,recvtype,comm)
MPI_ALLGATHER(sendbuf,sendcount,sendtype,recvbuf,
    recvcount,recvtype,comm,info)
```

MPI_Reduce

Applies a reduction operation on all tasks in the group and places the result in one task Table A.2.

```
MPI_Reduce(*sendbuf,*recvbuf,count,datatype,op,root,comm)
MPI_REDUCE(sendbuf,recvbuf,count,datatype,op,root,comm,ierr)
```

MPI reduction operations:

MPI_Allreduce

Applies a reduction operation and places the result in all tasks in the group. This is equivalent to an MPI_Reduce followed by an MPI_Bcast.

```
MPI_Allreduce(*sendbuf,*recvbuf,count,datatype,op,comm)
MPI_ALLREDUCE(sendbuf,recvbuf,count,datatype,op,comm,ierr)
```

Table A.2.   The predefined MPI reduction operations. Users can also define their own reduction functions by using the MPI_Op_create route.

MPI reduction Operation		C Data types	FORTRAN data types
MPI_MAX	maximum	integer, float	integer, real, complex
MPI_MIN	minimum	integer, float	integer, real, complex
MPI_SUM	sum	integer, float	integer, real, complex
MPI_PROD	product	integer, float	integer, real, complex
MPI_LAND	logical AND	integer	logical
MPI_BAND	bit-wise AND	integer, MPI_BYTE	integer, MPI_BYTE
MPI_LOR	logical OR	integer	logical
MPI_BOR	bit-wise OR	integer, MPI_BYTE	integer, MPI_BYTE
MPI_LXOR	logical XOR	integer	logical
MPI_BXOR	bit-wise XOR	integer, MPI_BYTE	integer, MPI_BYTE
MPI_MAXLOC	max value and location	float, double, long double	real, complex, double precision
MPI_MINLOC	min value and location	float, double, long double	real, complex, double precision

MPI_Reduce_scatter

First does an element-wise reduction on a vector across all tasks in the group. Next, the result vector is split into disjoint segments and distributed across the tasks. This is equivalent to an MPI_Reduce followed by an MPI_Scatter operation.

```
MPI_Reduce_scatter(*sendbuf,*recvbuf,recvcount,
    datatype,op,comm)
MPI_REDUCE_SCATTER(sendbuf,recvbuf,recvcount,
    datatype,op,comm,ierr)
```

MPI_Alltoall

Each task in a group performs a scatter operation, sending a distinct message to all the tasks in the group in order by index.

```
MPI_Alltoall(*sendbuf,sendcount,sendtype,*recvbuf,
    recvcnt,recvtype,comm)
MPI_ALLTOALL(sendbuf,sendcount,sendtype,recvbuf,
    recvcnt,recvtype,comm,ierr)
```

MPI_Scan

Performs a scan operation with respect to a reduction operation across a task group.

```
#include "mpi.h"
#include <stdio.h>
#define SIZE 4

int main(int argc, char *argv[]){
int numtasks, rank, sendcount, recvcount, source;
float sendbuf[SIZE][SIZE] = {
  { 1.0,  2.0,  3.0,  4.0},
  { 5.0,  6.0,  7.0,  8.0},
  { 9.0, 10.0, 11.0, 12.0},
  {13.0, 14.0, 15.0, 16.0}  };
float recvbuf[SIZE];

MPI_Init(&argc, &argv);
MPI_Comm_size(MPI_COMM_WORLD, &numtasks);
MPI_Comm_rank(MPI_COMM_WORLD, &rank);

if (numtasks == SIZE) {
    source = 1;
    sendcount = SIZE;
    recvcount = SIZE;
    MPI_Scatter(sendbuf, sendcount, MPI_FLOAT, recvbuf,
        recvcount, MPI_FLOAT, source, MPI_COMM_WORLD);

    printf("rank = %d Results: %f %f %f %f\n", rank,
        recvbuf[0], recvbuf[1], recvbuf[2], recvbuf[3]);
    }
else
    printf("Must specify %d processors, Terminating.\n",
        SIZE);
MPI_Finalize();
}
```

Fig. A.9. A simple example of collective communications in C, this code represents a scatter operation on the rows of an array.

```
MPI_Scan(*sendbuf,*recvbuf,count,datatype,op,comm)
MPI_SCAN(sendbuf,recvbuf,count,datatype,op,comm,ierr)
```

Examples: Figures A.10 and A.11 provide some simple examples of collective communications.

A.2. Examples of Using MPI

A.2.1. *"Hello" from all processes*

Compiling hello.c Fig. A.12

```
>mpicc -o hello hello.c
```

```
            program scatter
            include 'mpif.h'

            integer SIZE
            parameter (SIZE=4)
            integer numtasks, rank, sendcount
            integer recvcount, source, ierr
            real*4 sendbuf(SIZE, SIZE), recvbuf(SIZE)

c Fortran stores this array in column major order, so the
c scatter will actually scatter columns, not rows

            data sendbuf /  1.0,  2.0,  3.0,  4.0,
     &                      5.0,  6.0,  7.0,  8.0,
     &                      9.0, 10.0, 11.0, 12.0,
     &                     13.0, 14.0, 15.0, 16.0 /

            call MPI_INIT(ierr)
            call MPI_COMM_RANK(MPI_COMM_WORLD, rank, ierr)
            call MPI_COMM_SIZE(MPI_COMM_WORLD, numtasks, ierr)

            if (numtasks .eq. SIZE) then
                    source = 1
                    sendcount = SIZE
                    recvcount = SIZE
                    call MPI_SCATTER(sendbuf, sendcount, MPI_REAL,
     &                       recvbuf, recvcount, MPI_REAL, source,
     &                       MPI_COMM_WORLD, ierr)
                    print *, 'rank = ', rank, ' Results:', recvbuf
            else
                    print *, 'Must specify', SIZE,
     &                       ' processors. Terminating.'
            endif

            call MPI_FINALIZE(ierr)

            end
```

Fig. A.10. A simple example of collective communications in FORTRAN, this code represents a scatter operation on the rows of an array.

```
rank = 0 Results: 1.000000 2.000000 3.000000 4.000000
rank = 1 Results: 5.000000 6.000000 7.000000 8.000000
rank = 3 Results: 9.000000 10.000000 11.000000 12.000000
rank = 4 Results: 13.000000 14.000000 15.000000 16.000000
```

Fig. A.11. This is the output of the example given in.

Running hello.c Fig. A.12

```
>mpirun -np 4 hello
```

```
#include <stdio.h>
#include "mpi.h"

int main(int argc, char* argv[]) {
    int my_rank;                    /* rank of present process  */
    int p;                          /* number of processes      */
    MPI_Init(&argc, &argv);         /* initiate MPI             */
    MPI_Comm_rank(MPI_COMM_WORLD, &my_rank);
                                    /* find my rank in "comm"   */
    MPI_Comm_size(MPI_COMM_WORLD, &p);
                                    /* find "comm"'s size       */

    printf("Hello from process %d.\n", my_rank);

    MPI_Finalize();                 /* finish MPI               */
    return 0;
}
```

Fig. A.12. This example, hello.c, gives a "Hello" from all processes.

Output of hello.c Fig. A.12

```
Hello from process 0.
Hello from process 3.
Hello from process 2.
Hello from process 1.
```

A.2.2. *Integration*

Compiling integral.c Fig. A.13

```
>mpicc -o integral integral.c
```

Running integral.c Fig. A.13

```
>mpirun -np 4 integral
```

A.3. MPI Tools

Pallas

Pallas is a leading independent software company specializing in high performance computing. Pallas assisted many organizations in migrating from sequential to parallel computing. Customers of Pallas come from all fields: Hardware manufacturers, software vendors, as well as end users. Each of these has benefited from Pallas' unique experience in the development and tuning of parallel applications.

```
#include<stdio.h>
#include<math.h>
#include"mpi.h"

int main(int argc, char* argv[]){
    int my_rank;
    int p;
    int source;
    int dest;
    int tag=0;
    int i, n, N;            /* indices                  */
    float a, b;             /* local bounds of interval */
    float A, B;             /* global bounds of interval */
    float h;                /* mesh size                */
    float my_result;        /* partial integral         */
    float global_result;    /* global result            */
    MPI_Status status;      /* recv status              */

    MPI_Init(&artc, &argv);
    MPI_Comm_rank(MPI_COMM_WORLD, &my_rank);
    MPI_Comm_size(MPI_Comm_WORLD,&p);
    Read_Golbal_boundary(&A, &B); /* reads global bounds */
    h = (B-A)/N;            /* mesh size                */
    n = N/p;                /* meshes on each processor */
    a = A + my_rank*n*h;
    b = a+ n*h;             /* calculates the local bounds*/
    my_result = integral (f, a, b); /* perform integration*/

    /* Gathering the results to 0 */
    MPI_Reduce(&my_result,      /* sendbuf            */
            &global_result,     /* recvbuf            */
            1,                  /* # entries to send  */
            MPI_FLOAT,          /* datatype of entry  */
            MPI_SUM,            /* reduce operation   */
            0,                  /* send to my_rank=0  */
            MPI_COMM_WORLD);    /* communicator       */

    if (rank == 0)
      printf("The result is %f.\n", global_result);
                                /* print the result   */
    MPI_Finalize();
    return 0;
}
```

Fig. A.13.   The first part of integral.c, which performs a simple integral.

In the field of MPI development tools and implementations, Pallas contributions include:

- VAMPIR - MPI performance visualization and analysis
- VAMPIRtrace - MPI profiling instrumentation
- DIMEMAS - MPI application performance prediction
- MPI-2 - first industrial MPI-2 implementation in Nov. 1997

VAMPIR

VAMPIR is currently the most successful MPI tool product (see also "Supercomputer European Watch," July 1997), or check references at

(1) http://www.cs.utk.edu/~browne/perftools-review
(2) http://www.tc.cornell.edu/UserDoc/Software/PTools/vampir/

ScaLAPACK

The ScaLAPACK library includes a subset of LAPACK (Linear Algebra PACKage) routines redesigned for distributed-memory MIMD parallel computers. It is currently written in SPMD-type using explicit message passing for interprocessor communication. The goal is to have ScaLAPACK routines resemble their LAPACK equivalents as much as possible.

PGAPack

PGAPack is a general-purpose, data-structure-neutral, parallel genetic algorithm library. It is intended to provide most capabilities desired in a genetic algorithm library, in an integrated, seamless, and portable manner.

ARCH

ARCH is a C++-based object-oriented library of tools for parallel programming on computers using the MPI (message passing interface) communication library. Detailed technical information about ARCH is available as a Cornell Theory Center Technical Report (CTC95TR288).

http://www.tc.cornell.edu/Research/tech.rep.html

OOMPI

OOMPI is an object-oriented interface to the MPI-1 standard. While OOMPI remains faithful to all the MPI-1 functionality, it offers new object oriented abstractions that promise to expedite the MPI programming process by allowing programmers to take full advantage of C++ features.

http://www.cse.nd.edu/~lsc/research/oompi

XMPI: A Run/Debug GUI for MPI

XMPI is an X/Motif based graphical user interface for running and debugging MPI programs. It is implemented on top of LAM, an MPI cluster computing environment, but the interface is generally independent of LAM operating concepts. You write an MPI application in one or more MPI programs, tell XMPI about these programs and where they are to be run, and then snapshot the synchronization status of MPI processes throughout the application execution.

http://www.osc.edu/Lam/lam/xmpi.html

Aztec: An Iterative Sparse Linear Solver Package

Aztec is an iterative library that greatly simplifies the parallelization process when solving a sparse linear system of equations $Ax = b$ where A is a user supplied $n \times n$ sparse matrix, b is a user supplied vector of length n and x is a vector of length n to be computed. Aztec is intended as a software tool for users who want to avoid cumbersome parallel programming details but who have large sparse linear systems which require an efficiently utilized Parallel computing system.

http://www.cs.sandia.gov/HPCCIT/aztec.html

MPI Map

MPI Map, from Lawrence Livermore National Laboratory, lets programmers visualize MPI datatypes. It uses Tcl/Tk, and it runs on parallel computers that use the MPICH implementation of MPI. The tool lets you select one of MPI's type constructors (such as MPI Type vector or MPI Type struct) and enter the parameters to the constructor call. It then calls MPI to generate the new type, extracts the type map from the resulting structure, and presents a graphical display of the type map, showing the offset and basic type of each element.

http://www.llnl.gov/livcomp/mpimap/

STAR/MPI

STAR/MPI is a system to allow binding of MPI to a generic (STAR) interactive language. GCL/MPI is intended for easy-to-use master-slave distributed-memory architecture. It combines the feedback of an interactive language (the GCL or AKCL dialect of LISP) with the use of MPI to take advantage of networks of workstations.

ftp://ftp.ccs.neu.edu/pub/people/gene/starmpi/

Parallel Implementation of BLAS

The sB_BLAS package is a collection of parallel implementations of the level 3 Basic Linear Algebra Subprograms. All codes were written using MPI. A paper describing this work is also available.

http://www.cs.utexas.edu/users/rvdg/abstracts/sB_BLAS.html

BLACS (Basic Linear Algebra Communication Subprograms) for MPI

An "alpha test release" of the BLACS for MPI is available from the University of Tennessee, Knoxville. For more information contact R. Clint Whaley (rwhaley@cs.utk.edu).

NAG Parallel Library

The NAG Parallel Library is a library of numerical routines specifically produced for distributed-memory parallel computers. This library is available under MPI [or PVM] message-passing mechanisms. It also performs well on shared-memory computers whenever efficient implementations of MPI [or PVM] are available. It includes the following areas: Optimization, Dense linear algebra [including ScaLAPACK], Sparse linear algebra, Random number generators, Quadrature, Input/Output, data distribution, support/utility routines.

http://www.nag.co.uk/numeric/FM.html

DQS (Distributed Queuing System)

DQS now supports the launch of MPICH (Argonne/Miss State Version of MPI) jobs and is available by anonymous ftp.

ftp://ftp.scri.fsu.edu/pub/DQS

Interprocessor Collective Communication (iCC)

The interprocessor collective communication (iCC) research project started as a technique required to develop high performance implementations of the MPI collective communication calls.

http://www.cs.utexas.edu/users/rvdg/intercom/.

PETSc Scientific Computing Libraries

PETSc stands for "Portable Extensible Tools for scientific computing". It is a library of routines for both uni- and parallel-processor computing.

http://www.mcs.anl.gov/home/gropp/petsc.html.

MSG Toolkit

Message-passing tools for Structured Grid communications (MSG) is a MPI-based library intended to simplify coding of data exchange within FORTRAN 77 codes performing data transfers on distributed Cartesian grids. More information, the source code, and the user's guide are available at the following site:

http://www.cerca.umontreal.ca/malevsky/MSG/MSG.html.

Para++

The Para++ project provides a C++ interface to the MPI and PVM message passing libraries. Their approach is to overload input and output operators to do communication. Communication looks like standard C++ I/O.

http://www.loria.fr/para++/parapp.html.

Amelia Vector Template Library

The Amelia Vector Template Library (AVTL) is a polymorphic collection library for distributed-memory parallel computers. It is based on ideas from the Standard Template Library (STL) and uses MPI for communication.

ftp://riacs.edu/pub/Excalibur/avtl.html.

Parallel FFTW

FFTW, a high-performance, portable C library for performing FFTs in one or more dimensions, includes parallel, multi-dimensional FFT routines for MPI. The transforms operate in-place for arbitrary-size arrays (of any dimensionality greater than one) distributed across any number of processors. It is free for non-commercial use.

http://www.fftw.org/.

Cononical Classes for Concurrency Control

The Cononical Classes for Concurrency Control library contains a set of C++ classes that implement a variety of synchronization and data transmission paradigms. It currently supports both Intel's NX and MPI.

http://dino.ph.utexas.edu/~furnish/c4.

MPI Cubix

MPI Cubix is an I/O library for MPI applications. The semantics and language binding reflect POSIX in its sequential aspects and MPI in its parallel aspects. The library is built on a few POSIX I/O functions and each of the POSIX-like Cubix functions translate directly to a POSIX operation on a file system somewhere in the parallel computer. The library is also built on MPI and is therefore portable to any computer that supports both MPI and POSIX.

http://www.osc.edu/Lam/mpi/mpicubix.html.

MPIX Intercommunicator Extensions

The MPIX Intercommunicator Extension library contains a set of extensions to MPI that allow many functions that previously only worked with intracommunicators to work with intercommunicators. Extensions include support for additional intercommunciator construction operations and intercommunicator collective operations.

http://www.erc.msstate.edu/mpi/mpix.html.

mpC

mpC, developed and implemented on the top of MPI, is a programming environment facilitating and supporting efficiently portable modular parallel programming. mpC does not compete with MPI, but tries to strengthen its advantages (portable modular programming) and to weaken its disadvantages (a low level of parallel primitives and difficulties with efficient portability; efficient portability means that an application running efficiently on a particular multi-processor will run efficiently after porting to other multi-processors). In fact, users can consider mpC as a tool facilitating the development of complex and/or efficiently portable MPI applications.

http://www.ispras.ru/~mpc/.

MPIRUN

Sam Fineberg is working on support for running multidisciplinary codes using MPI which he calls MPIRUN. You can retrieve the MPIRUN software from,

http://lovelace.nas.nasa.gov/Parallel/People/fineberg_homepage.html.

A.4. Complete List of MPI Functions

Constants	MPI_Initialized
MPI_File_iwrite_shared	MPIO_Wait
MPI_Info_set	MPI_File_read_all
MPIO_Request_c2f	MPI_Int2handle
MPI_File_open	MPI_Abort
MPI_Init	MPI_File_read_all_begin
MPIO_Request_f2c	MPI_Intercomm_create
MPI_File_preallocate	MPI_Address
MPI_Init_thread	MPI_File_read_all_end
MPIO_Test	MPI_Intercomm_merge
MPI_File_read	MPI_Allgather

MPI_File_read_at
MPI_Iprobe
MPI_Allgatherv
MPI_File_read_at_all
MPI_Irecv
MPI_Allreduce
MPI_File_read_at_all_begin
MPI_Irsend
MPI_Alltoall
MPI_File_read_at_all_end
MPI_Isend
MPI_Alltoallv
MPI_File_read_ordered
MPI_Issend
MPI_Attr_delete
MPI_File_read_ordered_begin
MPI_Keyval_create
MPI_Attr_get
MPI_File_read_ordered_end
MPI_Keyval_free
MPI_Attr_put
MPI_File_read_shared
MPI_NULL_COPY_FN
MPI_Barrier
MPI_File_seek
MPI_NULL_DELETE_FN
MPI_Bcast
MPI_File_seek_shared
MPI_Op_create
MPI_Bsend
MPI_File_set_atomicity
MPI_Op_free
MPI_Bsend_init
MPI_File_set_errhandler
MPI_Pack
MPI_Buffer_attach
MPI_File_set_info
MPI_Pack_size
MPI_Buffer_detach

MPI_File_set_size
MPI_Pcontrol
MPI_CHAR
MPI_File_set_view
MPI_Probe
MPI_Cancel
MPI_File_sync
MPI_Recv
MPI_Cart_coords
MPI_File_write
MPI_Recv_init
MPI_Cart_create
MPI_File_write_all
MPI_Reduce
MPI_Cart_get
MPI_File_write_all_begin
MPI_Reduce_scatter
MPI_Cart_map
MPI_File_write_all_end
MPI_Request_c2f
MPI_Cart_rank
MPI_File_write_at
MPI_Request_free
MPI_Cart_shift
MPI_File_write_at_all
MPI_Rsend
MPI_Cart_sub
MPI_File_write_at_all_begin
MPI_Rsend_init
MPI_Cartdim_get
MPI_File_write_at_all_end
MPI_Scan
MPI_Comm_compare
MPI_File_write_ordered
MPI_Scatter
MPI_Comm_create
MPI_File_write_ordered_begin
MPI_Scatterv
MPI_Comm_dup

MPI_File_write_ordered_end

MPI_Send

MPI_Comm_free

MPI_File_write_shared

MPI_Send_init

MPI_Comm_get_name

MPI_Finalize

MPI_Sendrecv

MPI_Comm_group

MPI_Finalized

MPI_Sendrecv_replace

MPI_Comm_rank

MPI_Gather

MPI_Ssend

MPI_Comm_remote_group

MPI_Gatherv

MPI_Ssend_init

MPI_Comm_remote_size

MPI_Get_count

MPI_Start

MPI_Comm_set_name

MPI_Get_elements

MPI_Startall

MPI_Comm_size

MPI_Get_processor_name

MPI_Status_c2f

MPI_Comm_split

MPI_Get_version

MPI_Status_set_cancelled

MPI_Comm_test_inter

MPI_Graph_create

MPI_Status_set_elements

MPI_DUP_FN

MPI_Graph_get

MPI_Test

MPI_Dims_create

MPI_Graph_map

MPI_Test_cancelled

MPI_Errhandler_create

MPI_Graph_neighbors

MPI_Testall

MPI_Errhandler_free

MPI_Graph_neighbors_count

MPI_Testany

MPI_Errhandler_get

MPI_Graphdims_get

MPI_Testsome

MPI_Errhandler_set

MPI_Group_compare

MPI_Topo_test

MPI_Error_class

MPI_Group_difference

MPI_Type_commit

MPI_Error_string

MPI_Group_excl

MPI_Type_contiguous

MPI_File_c2f

MPI_Group_free

MPI_Type_create_darray

MPI_File_close

MPI_Group_incl

MPI_Type_create_subarray

MPI_File_delete

MPI_Group_intersection

MPI_Type_extent

MPI_File_f2c

MPI_Group_range_excl

MPI_Type_free

MPI_File_get_amode

MPI_Group_range_incl

MPI_Type_get_contents

MPI_File_get_atomicity

MPI_Group_rank

MPI_Type_get_envelope

MPI_File_get_byte_offset

MPI_Group_size

MPI_Type_hvector

MPI_File_get_errhandler

MPI_Group_translate_ranks	MPI_File_get_view
MPI_Type_lb	MPI_Info_f2c
MPI_File_get_group	MPI_Waitall
MPI_Group_union	MPI_File_iread
MPI_Type_size	MPI_Info_free
MPI_File_get_info	MPI_Waitany
MPI_Ibsend	MPI_File_iread_at
MPI_Type_struct	MPI_Info_get
MPI_File_get_position	MPI_Waitsome
MPI_Info_c2f	MPI_File_iread_shared
MPI_Type_ub	MPI_Info_get_nkeys
MPI_File_get_position_shared	MPI_Wtick
MPI_Info_create	MPI_File_iwrite
MPI_Type_vector	MPI_Info_get_nthkey
MPI_File_get_size	MPI_Wtime
MPI_Info_delete	MPI_File_iwrite_at
MPI_Unpack	MPI_Info_get_valuelen
MPI_File_get_type_extent	MPI_File_iwrite_shared
MPI_Info_dup	MPI_Info_set
MPI_Wait	

APPENDIX B

OPENMP

B.1. Introduction to OpenMP

OpenMP stands for Open Multi-Processing, which is the *defacto* standard API for writing shared-memory parallel application. The general idea of OpenMP is multi-threading by fork-join model (Fig. B.1): All OpenMP programs begin as a single process called the master thread. When the master thread reaches the parallel region, it creates multiple threads to execute the parallel codes enclosed in the parallel region. When the threads complete the parallel region, they synchronize and terminate, leaving only the master thread.

B.1.1. *The brief history of OpenMP*

In the early 1990s, vendors supplied similar, directive-based programming extensions for their own shared-memory computers. These extensions offered preprocess directives where the user specified which loops were to be parallelized within a serial program and then the compiler would be responsible for automatically paralleling such loops across the processors. All these implementations were all functionally similar, but were diverging.

The first attempt to standardize the shared-memory API was the draft for ANSI X3H5 in 1994. Unfortunately, it was never adopted. In October 1997, the OpenMP Architecture Review Board (ARB) published its first API specifications, OpenMP for Fortran 1.0. In 1998 they released the C/C++ standard. The version 2.0 of the Fortran specifications was published in 2000 and version 2.0 of the C/C++ specifications was released in 2002. Version 2.5 is a combined C/C++/Fortran specification that was released in 2005. Version 3.0, released in May 2008, is the current version of the API specifications.

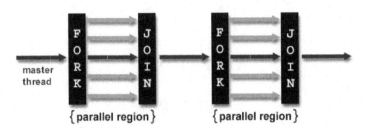

Fig. B.1.   Fork-join model used in OpenMP.

B.2. Memory Model of OpenMP

OpenMP provides a relaxed-consistency, shared-memory model. All OpenMP threads have access to a place to store and to retrieve variables, called the memory. In addition, each thread is allowed to have its own temporary view of the memory. The temporary view of memory for each thread is not a required part of the OpenMP memory model, but can represent any kind of intervening structure, such as machine registers, cache, or other local storage, between the thread and the memory. The temporary view of memory allows the thread to cache variables and thereby to avoid going to memory for every reference to a variable. Each thread also has access to another type of memory that must not be accessed by other threads, called *threadprivate* memory.

B.3. OpenMP Directives

OpenMP is directive based. Most parallelism is specified through the use of compiler directives which are embedded in C/C++ or FORTRAN. A directive in C/C++ has the following format:

`#pragma omp <directive-name> [clause,...]`

Example

`#pragma omp parallel num_threads(4)`

B.3.1. *Parallel region construct*

The directive to create threads is,

`#pragma omp parallel [clause [, ]clause] ...]`

where clause is one of the following:

```
if(scalar-expression)
num_threads(integer-expression)
default(shared | none)
private(list)
firstprivate(list)
shared(list)
copyin(list)
reduction(operator: list).
```

B.3.2. *Work-sharing constructs*

A work-sharing construct distributes the execution of the associated region among the threads encountering it. However, it does not launch new threads by itself. It should be used within the parallel region or combined with the parallel region constructs.

The directive for the loop construct is as follows:

```
#pragma omp for [clause[[,] clause] ...]
```

where the clause is one of the following:

```
private(list)
firstprivate(list)
lasstprivate(list)
reduction(operator: list)
schedule(kind[, chunck_size])
collapse(n)
ordered
nowait
```

B.3.3. *Directive clauses*

The `private` clause declares variables in its list to be private to each thread. For a private variable, a new object of the same type is declared once for each thread in the team and all reference to the original object are replaced with references to the new object. Hence, variables declared private should be assumed to be uninitialized for each thread.

The `firstprivate` clause has the same behavior of `private` but it automatically initializes the variables in its list according to their original values.

The `lastprivate` clause does what `private` does and copies the variable from the last loop iteration to the original variable object.

The `shared` clause declares variables in its list to be shared among all threads.

The `default` clause allows user to specify a default scope for all variables within a parallel region to be either shared or not.

The `copyin` clause provides a means for assigning the same value to `threadprivate` variables for all threads in the team.

The `reduction` clause performs a reduction on the variables that appear in its list. A private copy for each list variable is created for each thread. At the end of the reduction, the reduction variable is applied to all private copies of the shared variable, and the final result is written to the global shared variable.

B.4. Synchronization

Before we discuss the synchronization, let us consider a simple example where two processors try to do a read/update on same variable.

```
x = 0;
#pragma omp parallel shared(x)
{
    x = x + 1;
}
```

One possible execution sequence is as follows:

1. Thread 1 loads the value of x into register A.
2. Thread 2 loads the value of x into register A.
3. Thread 1 adds 1 to register A.
4. Thread 2 adds 1 to register A.
5. Thread 1 stores register A at location x.
6. Thread 2 stores register A at location x.

The result of x will be 1, not 2 as it should be. To avoid situation like this, x should be synchronized between two processors.

Critical

The `critical` directive specifies a region of code that must be executed by only one thread at a time. An optional name may be used to identify the `critical` construct. All `critical` without a name are considered to have the same unspecified name. So if there exist two or more independent blocks of `critical` procedure in the same parallel region, it is important to specify them with different name so that these blocks can be executed in parallel.

```
#pragma omp critical [name]
    structured_block
```

atomic

The `atomic` directive ensures that a specific storage location is updated atomically, rather than exposing it to the possibility of multiple, simultaneous writings (race condition). The `atomic` directive applies only to the statement immediately following it and only the variable being updated is protected by `atomic`. The difference between `atomic` and `critical` is that `atomic` operations can be executed in parallel when updating different element while `critical` blocks are guaranteed to be serial.

```
#pragma omp atomic
    statement_expression
```

barrier

The `barrier` directive synchronizes all threads in the team. When a `barrier` directive is reached, a thread will wait at that point until all other threads have reached that barrier. All threads then resume executing in parallel the code that follows the barrier.

```
#pragma omp barrier
```

flush

The `flush` directive identifies a synchronization point at which the implementation must provide a consistent view of memory. Thread-visible variables are written back to memory at this point.

```
#pragma omp flush (list)
```

B.5. Runtime Library Routines

The OpenMP standard defines an API for library calls that perform a variety of functions: Query the number of threads/processors, set number of threads to use; general purpose locking routines (semaphores); portable wall clock timing routines; set execution environment functions: Nested parallelism, dynamic adjustment of threads. For C/C++, it may be necessary to specify the include file "omp.h".

B.5.1. *Execution environment routines*

omp_set_num_threads

This sets the number of threads that will be used in the next parallel region. It must be a positive integer.

void omp_set_num_threads(int num_threads).

omp_get_num_threads

Returns the number of threads that are currently in the team executing the parallel region from which it is called.

int omp_get_num_threads(void).

omp_get_max_threads

This returns the maximum value that can be returned by a call to the OMP_GET_NUM_THREADS function.

int omp_get_max_threads(void).

omp_get_thread_num

Returns the thread number of the thread, within the team, making this call. This number will be between 0 and OMP_GET_NUM_THREADS - 1. The master thread of the team is thread 0.

int omp_get_thread_num(void).

omp_in_parallel

To determine if the section code which is executing is parallel or not.

int omp_in_parallel(void).

B.5.2. *Lock routines*

The OpenMP runtime library includes a set of general-purpose lock routines that can be used for synchronization. These general-purpose lock routines operate on OpenMP locks that are represented by OpenMP lock variables. An OpenMP lock variable must be accessed only through the routines described in this section; programs that otherwise access OpenMP lock variables are non-conforming.

The OpenMP lock routines access a lock variable in such a way that they always read and update the most current value of the lock variable. The lock routines include a flush with no list; the read and update to the lock variable must be implemented as if they are atomic with the flush. Therefore, it is not necessary for an OpenMP program to include explicit flush directives to ensure that the lock variable's value is consistent among different tasks.

omp_init_lock

This subroutine initializes a lock associated with the lock variable.

`void omp_init_lock(omp_lock_t *lock).`

omp_destroy_lock

This subroutine disassociates the given lock variable from any locks.

`void omp_destroy_lock(omp_lock_t *lock).`

omp_set_lock

This subroutine forces the executing thread to wait until the specified lock is available. A thread is granted ownership of a lock when it becomes available.

`void omp_set_lock(omp_lock_t *lock).`

omp_unset_lock

This subroutine releases the lock from the executing subroutine.

`void omp_unset_lock(omp_lock_t *lock).`

omp_test_lock

This subroutine attempts to set a lock, but does not block if the lock is unavailable.

`int omp_test_lock(omp_lock_t *lock)`

B.5.3. *Timing routines*

omp_get_wtime

This routine returns elapsed wall clock time in seconds.

`double omp_get_wtime(void).`

omp_get_wtick

This routine returns the precision of the timer used by `omp_get_wtime`.

`double omp_get_wtick(void)`.

B.6. Examples of Using OpenMP

Similar to the MPI chapter, here also we provide two simple examples (Figs. B.2 and B.3) of using OpenMP.

B.6.1. *"Hello" from all threads*

```
#include"omp.h"

#include<stdio.h>

int main(int argc, char* argv[])

{

#pragma omp parallel num_threads(4)

    {

        int ID = omp_get_thread_num();

        int N  = omp_get_num_threads();

        printf("Hello from %d of %d.\n", ID, N);

    }

    return 0;

}
```

Fig. B.2.   A simple "Hello" program from all threads.

B.6.2. *Calculating π by integration*

```c
#include"omp.h"

#include<stdio.h>

static long num_step = 100000;

double step;

#define NUM_THREADS 2

int main(int argc, char* argv[])

{

    int i;

    double x, pi, sum = 0.0;

    step = 1.0/(double)num_steps;

    omp_set_num_threads(NUM_THREADS);

#pragma omp parallel for private(x) reduction(+:sum)

    for (i = 0;i < num_steps; i++) /* i private by
default */

    {

        x = (i + 0.5) * step;

        sum = sum + 4.0/(1.0 + x * x);

    }

    pi = step * sum;

    printf("Computed pi = %f\n", pi);

    return 0;

}
```

Fig. B.3. OpenMP example for computing pi by integration.

B.7. The Future

MPI and OpenMP have served the parallel computing communities' needs of portability and convenience of message passing over the past nearly 20 years. Like everything else, they are approaching the second half of their useful lifetime after passing the peaks. Parallel computers are monsters of millions of connected processing cores with fancy network topologies and protocols. A new programming model and, certainly, a new tool to enable their coordination or communication, are highly desired. Many attempts, just like those of pre-MPI era, are starting the bear fruits and, expectedly, a new technique similar to MPI and OpenMP will emerge.

APPENDIX C

PROJECTS

Project C.1 Watts and Flops of Supercomputers

From the latest (Nov. 2011) Top500 and Green500 lists, one can find the ordered lists of Top500 and Green500 supercomputers. The Top500 supercomputers are ranked in terms of the sustained floating-point operations Rmax in unit of Gflops while the Green500 supercomputers are ranked in terms of their power efficiency measured in terms of Mflops per Watt of electricity. Assume the Top500 computers are marked with $L = 1, 2, \ldots, 500$ where $L = 1$ is assigned to the computer with the highest Rmax while the Green500 computers are marked with $P = 1, 2, \ldots, 500$ where $P = 1$ is assigned to the computer with the highest performance-to-power ratio.

Please name all supercomputers whose values satisfy $P + L \leq 50$ and construct a table to list the features of their processors (CPUs and GPUs if there are any), their number of processing cores, and their interconnection networks.

Project C.2 Review of Supercomputers

Supercomputers have been going through active development for the last 50 years. Write an essay not less than 10 pages to document such development. In the essay, describe the evolution of the following aspects:

(1) key architecture including the processing and network units.
(2) system performance.
(3) programming models.
(4) representative applications.
(5) major breakthroughs and bottleneck.

Project C.3 Top500 and BlueGene Supercomputers

According to www.top500.org in 2010, many of the world's fastest supercomputers belong to the BlueGene family designed and constructed by IBM. We have to do the following:

(1) Define the main architectural components of the IBM BlueGene/P or BlueGene/Q supercomputers.
(2) Explain how world's top500 supercomputers are selected
    (i) Describe benchmark used?
    (ii) Describe the pros and cons of such benchmarking system.
    (iii) How would you improve the benchmarking system?
    (iv) Are there any other benchmarks?
(3) Design an algorithm to solve dense linear algebraic equation on BlueGene/P or BlueGene/Q.

Write an essay to compare five of the top ten supercomputers from the latest top500.org list. For example, the top ten supercomputers of the June 2011 list include:

(1) Fujitsu K computer.
(2) NDUT Tianhe-1A.
(3) Cray XT5-HE.
(4) Dawning TC3600 Cluster.
(5) HP Cluster Platform 3000SL.
(6) Cray XE6.
(7) SGI Altix ICE.
(8) Cray XE6.
(9) Bull Bullx supernode S6010/6030.
(10) Roadrunner IBM BladeCenter.

The five computers that are selected should be as diverse as possible. The essay, at least 10 pages in length, must contain the following aspects:

(1) key architecture including the processing and network units.
(2) system performance.
(3) programming models.
(4) major advantageous features for each computer.

In each of the categories, create a rank order for the computers you select.

Project C.4 Say Hello in Order

When running a "Hello from Processor X" program on P processors, you usually see output on your screen ordered randomly. It's not in the order you desired, e.g., "Hello from Processor 1", then "Hello from Processor 2", etc.

Write a parallel program to enforce the order, i.e., whenever you run it, you always get the processors to say hello to you in order. You may test it for a system with at least 13 processors.

Project C.5 Broadcast on Torus

Broadcast 10 floating-point numbers from any point to all 6000 nodes of a 3D torus $10 \times 20 \times 30$. Please design an algorithm to quickly complete the broadcast. Please report the total number of steps "N" needed to complete the broadcast. Additionally, make a table to show the number of links that are involved in each of the "N" steps for the broadcast.

Project C.6 Competing with MPI on Broadcast, Scatter, etc

MPI implementation has a dozen or so collective operation functions such as broadcast, gather, scatter, all-gather, etc. Describe the algorithms of implementation three functions: Broadcast, scatter, and all-gather.

Implement such functions on your chosen computer using only MPI_Isend() and MPI_Irecv() and a few other supporting (non-data-motion) functions and other light-weight basic single-sided communication functions. Then, compare the performance of your implementation with that of those provided by MPI. You need to vary the size of the messages (number of floating-point numbers to move) and the size of the computer involved to study the dependence of the differences on the message and machine sizes. For example, if you want to perform collective operations on $N = 10^2, 10^3, 10^4$ floating numbers from $P = 2^2, 2^3, 2^4, 2^5$ processing cores.

Project C.7 Simple Matrix Multiplication

Create two $N \times N$ square matrices with random elements whose values are in $[-1, 1]$ and compute their product by using P processors. The program should be general for reasonable values of N and P. Plot the speedup curves at $N = 1000, 2000, 4000$ and $P = 1, 2, 4, 8$ processors.

Project C.8 Matrix Multiplication on 4D Torus

Design an algorithm to multiply two large square matrices A and B on a 4D torus networked supercomputer with $15 \times 15 \times 16 \times 9 = 32,400$ nodes. You may consider A and B having $8 \times 8 \times (180 \times 180)$ elements, i.e., your matrices are of the size 1440×1440. Further, we assume the time to perform one floating-point operation (adding or multiplying a pair of floating-point numbers) is 1 unit (of time). We also assume transmitting one floating-point number between any adjacent nodes takes 1 unit (of time). Estimate the shortest time to complete the matrix multiplication on the computer.

Project C.9 Matrix Multiplication and PAT

There are several parallel matrix multiplication algorithms for distributed-memory parallel computers with various communication networks. In this project, we will use one hypothetical parallel computer with $P = 8 \times 8 \times 8 = 512$ processing cores on a 3D mesh network. Now, we hope to design two algorithms, Ring method (aka Canon method) and the BMR method, to multiply two matrices of the size 1024×1024. We also assume the time to take each core to perform one floating-point operation is "t_0" and the time to communicate one floating-point number from one core to any one of its nearest neighbors is "t_c". To simplify the problem, the time for communicating to the next to the nearest neighbors is "$2t_c$", to the next nearest neighbors is "$3t_c$", etc.

The project requires you to draw two Parallel Activity Trace (PAT) graphs, one for each algorithm, to illustrate the activities in the cores during the entire process of completing the matrix multiplication. A PAT is aligned graphs of time series from all the processors. Each time series records the activities of each processor during the matrix multiplication: sequential operations, message send, message receive, or idle. Hopefully the PAT is reasonably in scale.

Project C.10 Matrix Inversion

For a typical distributed-memory parallel computer, memory contents among all processors are shared through message passing. One standard for such sharing is called MPI, i.e. message passing interface. We have to do the following:

(1) Define the main function groups in MPI 2.0.
(2) If a distributed-memory parallel computer is used to invert a dense square matrix:

   (i) Design an algorithm for the matrix inversion.
   (ii) Analyze the performance of the algorithm.
   (iii) Write a parallel program for implementing the algorithm for inverting a 1024×1024 matrix on a parallel computer with $2^0, 2^1, 2^3$ and 2^4 processors.

Project C.11 Simple Analysis of an iBT Network

Starting from a 3D torus $30 \times 30 \times 36$, you add links that bypass 6 or 12 hops systematically in all three dimensions and for all nodes. This will get you a new network that we call Interlacing Bypass Torus (IBT) with node degree 8, i.e., each node has, and only has, 8 links. You may also construct a 4D torus network with the same number of nodes. You will have many options to assemble the 4D torus. We request you make it as $15 \times 15 \times 16 \times 9$.

Please compute, for both networks, (1) the average node-to-node distances and (2) the diameters, measured in hops.

Project C.12 Compute Eigenvalues of Adjacency Matrices of Networks

One can form many networks topologies to connect $2^6 \times 2^6 = 2^{12}$ nodes. For our project, we form the following three topologies:

(1) A 2D torus of $2^6 \times 2^6$
(2) A 3D mesh of $2^3 \times 2^3 \times 2^4$
(3) A hypercube of 12 dimension, 2^{12}

Now, construct the adjacency matrices for each of the three topologies. The adjacency matrix $S_{N \times N}$ for an N-node system is defined as an $N \times N$ matrix whose elements S_{ij} is the distance between nodes i and j measured in hops. Next, compute the eigenvalues for each of the three adjacency matrices.

Project C.13 Mapping Wave Equation to Torus

Solve a wave equation defined on a 2D grid of $M \times N$ mesh points on a parallel computer with $P \times Q \times R$ processors. These processors are arranged on 3D grid with the nearest neighbor communication links at latency a and bandwidth b per link. We assume that the processors at the ends of x-, y-, and z-directions are connected, i.e. the computer network is a 3D torus. We further assume that the number of mesh points is much higher than that of processors in the system. We have to do the following:

(1) Map the 2D computational domain to the 3D computer network.
(2) Argue that the mapping is near optimal for message passing.
(3) Estimate the speedup that you may get for the mapping.
(4) Write a program to solve the wave equation on an emulated distributed-memory MIMD parallel computer of $2 \times 2 \times 2$ for the equation problem on 128×128 mesh.

Project C.14 Load Balance in 3D Mesh

Solve 2D wave equation on a 3D hypothetical parallel computer. The wave equation is defined on a 2D square domain with 256 mesh-points in each dimension. The equation is,

$$
\begin{cases}
\dfrac{\partial^2 u}{\partial t^2} = c^2 \left(\dfrac{\partial^2 u}{\partial x^2} + \dfrac{\partial^2 u}{\partial y^2} \right) \\[2mm]
u(t = 0) = u_0 \\
u'(t = 0) = v_0 \\
u(x = 0) = u(x = L) \\
u(y = 0) = u(y = L)
\end{cases}
$$

where c, u_0, v_0, L are constants.

Now, the computer consists of 64 processors with the same speed (1 Gflops each) and these processors are placed on a 3D mesh $4 \times 4 \times 4$. Each link connecting any two nearest neighboring processors is capable of 1 Gbps communication bandwidth. Assume we use a finite difference method to solve the problem. We have to do the following:

(1) Design an algorithm to place the 256×256 mesh points to the 64 processors with optimal utilization and load balance of the CPUs and communication links.
(2) Compute the expected load imbalance ratio for CPUs resulting from the method.
(3) Compute the expected load imbalance ratio among all links resulting from the method.
(4) Repeat the above three steps if each one of the top layer of 16 processors is four times faster than each one of the processors on the second and third layers.

Project C.15 Wave Equation and PAT

Please write a parallel program on a 3D or higher dimensional mesh networked parallel computer to solve 2D wave equation by finite difference method. The wave equation is defined on a two-dimensional square domain with 256 mesh-points in each dimension. The equation is

$$
\begin{cases}
u_{tt} = c^2 (u_{xx} + u_{yy}) \\
u(t = 0) = u_0, \ u'(t = 0) = v_0 \\
u(x = 0) = u(x = L), \ u(y = 0) = u(y = L) \\
\text{where } c, \ u_o, \ v_0, \ L \text{ are constants}
\end{cases}
$$

Assume you solve the above on a computer with $4 \times 4 \times 4 = 64$ processing cores. Obviously, the internal network for the cores in a processor is unknown and we do not care. If you only have access to a Beowulf cluster, you may use it with any network available.

Please draw the PAT (Parallel Activity Traces) for running your program for 10 time steps with the actual computing and communication data from your calculations. You should include the serial (local) computing, message passing, and idle of each of the participating processing cores.

Project C.16 Computing Coulomb's Forces

In 3D, N particles are assigned electric charges with values taking random numbers in $\{\pm1, \pm2, \pm3\}$. In other words, the charge of the each particle is one of the six random numbers with equal probability. These particles are fixed (for convenience) at sites whose coordinates are random numbers in a $10 \times 10 \times 10$ cube, i.e. the Cartesian coordinates of these N particles are any triplet of random numbers in $[0, 10]$. These particles interact under the Coulomb's law:

$$F_{ij} = \frac{q_i q_j}{r_{ij}^2},$$

where F_{ij} is the force on particle i by j (with distance r_{ij}) with charges q_i and q_j.

Write a program to compute the forces on each of the N particles by P processors. Make the following plot: A set of speedup curves at different values of N.

(1) For $N = 1000$, collect timing results for $P = 2^0, 2^1, 2^2, 2^3, 2^4, 2^5$.
(2) For $N = 2000$, collect timing results for $P = 2^0, 2^1, 2^2, 2^3, 2^4, 2^5$.
(3) For $N = 5000$, collect timing results for $P = 2^0, 2^1, 2^2, 2^3, 2^4, 2^5$.

Note: When collecting timing results, minimize I/O effect.

Project C.17 Timing Model for MD

In molecular dynamics (MD) simulation of proteins or other bio-molecules, one must compute the bonded and non-bonded forces of each atom before solving the Newton's equation of motion, for each time step. Design a simple timing model for MD calculation. For example, first, construct a table to estimate the number of operations needed to compute each of the force terms and solution of the ODEs for $P = 1$ processor. Next, you divide the task into P sub-tasks and assign to P individual processors or nodes. Assume the processor speed f, inter-node communication latency t_0 and communication bandwidth w and compute the time to finish the MD simulation on such a system with P processors.

If you solve this MD problem on a specific computer (In 2011, a popular computer is BlueGene/P), please look up the relevant computer parameters (including processor speed and network latency and bandwidth) to estimate the time needed, for each time step, for the major calculation components. With such computer specifications and our timing model, we estimate the parallel efficiency for a molecule with $N = 100,000$ atoms to be modeled by the given supercomputer with $P = 2^{10}, 2^{12}, 2^{14}, 2^{16}$ processors.

Project C.18 Minimizing Lennard-Jones Potential

In three dimensions, N particles are fixed at sites that have random coordinates in a $10 \times 10 \times 10$ cube, i.e. any triplet of random numbers in $[0, 10]$ can be the coordinates of a particle, initially. These particles interact under the so-called Lennard-Jones potential,

$$V_{ij} = \frac{1}{r_{ij}^{12}} - \frac{2}{r_{ij}^6}$$

where V_{ij} is the pair-wise potential between particles i and j with distance r_{ij}.

(1) Write a serial program to minimize the total energy of the particle system for $N = 100$ by using simulated annealing method (or any optimization method that you prefer).

(2) For $P = 2, 4$, and 8, write a parallel program to minimize the energy. The minimizations should terminate at a similar final energy that can be obtained by $P = 1$ as in (1) above (Results within five percent of relative errors are considered similar.) We need to use the following two methods to decompose the problem:

   (a) Particle decomposition.
   (b) Spatial decomposition.

(3) Report the timing results and speedup curve for both decompositions and comment on their relative efficiency.

Project C.19 Install and Profile CP2K

CP2K is a program to perform atomistic and molecular simulations of solid state, liquid, molecular, and biological systems. It provides a general framework for different methods such as, e.g., density functional theory (DFT) using a mixed Gaussian and plane waves approach (GPW) and classical pair and many-body potentials. CP2K is freely available under the GPL license. It is written in FORTRAN-95 and can be run efficiently in parallel.

Download the latest source code and compile the program for a parallel computer you can access. Run the program on that computer with 8, 16, and 32 processing cores for 100 time steps. You may use the default parameters or change them to a reasonable set that you can explain. Please build a table to show the wall clock times on all processing cores for their local calculations (accumulated over the 100 time steps). More desirably, profile the program.

Project C.20 Install and Profile CPMD

CPMD is a well parallelized, plane wave and pseudo-potential implementation of Density Functional Theory, developed especially for ab-inito molecular dynamics. It was developed at IBM Zurich Research Laboratory. The key features of CPMD include

- works with norm conserving or ultra-soft pseudo-potentials
- LDA, LSD and the most popular gradient correction schemes; free energy density functional implementation
- isolated systems and system with periodic boundary conditions; k-points
- molecular and crystal symmetry
- wavefunction optimization: direct minimization and diagonalization
- geometry optimization: local optimization and simulated annealing
- molecular dynamics: constant energy, constant temperature and constant pressure
- path integral MD
- response functions
- excited states
- many electronic properties
- time-dependent DFT (excitations, molecular dynamics in excited states)
- coarse-grained non-Markovian meta dynamics

Download the latest source code and compile the program for a parallel computer you can access. Run the program on that computer with 8, 16, and

32 processing cores for 100 time steps. You may use the default parameters or change them to a reasonable set that you can explain. Please build a table to show the wall clock times on all processing cores for their local calculations (accumulated over the 100 time steps). More desirably, profile the program.

Project C.21 Install and Profile NAMD

NAMD is a parallel molecular dynamics code designed for high-performance simulation of large bio-molecular systems. Based on Charm++ parallel objects (in 2010), NAMD scales to hundreds of processors on high-end parallel platforms and tens of processors on commodity clusters using gigabit Ethernet. One can build NAMD from the source code or download the binary executable for a wide variety of platforms.

Download the latest source code and compile the program for a parallel computer you can access. Run the program on that computer with 8, 16, and 32 processing cores for 100 time steps. You may use the default parameters or change them to a reasonable set that you can explain. Please build a table to show the wall clock times on all processing cores for their local calculations (accumulated over the 100 time steps). More desirably, profile the program.

Project C.22 FFT on Beowulf

Fast Fourier Transform (FFT) is important for many problems in computational science and engineering and its parallelization is very difficult to scale for large distributed-memory systems.

Suppose you are required to perform FFT efficiently on a Beowulf system with two layers of switches. Hypothetically, the Beowulf system has 480 processors that are divided into 32 groups with 15 each. The 15-processor group is connected to a Gigabit switch with 16 ports, one of which is connected to the "master" 10-Gigabit switch that has 32 ports. We also assume the latencies for the Gigabit and 10-Gigabit switches 50 and 20 microseconds respectively.

(1) Draw a diagram to illustrate the Beowulf system.
(2) Design the algorithm for FFT on such Beowulf system.
(3) Estimate the speedup for the algorithm.

Project C.23 FFT on BlueGene/Q

If you perform a Fourier transform for a 3D function $f(x, y, z)$ on BlueGene/Q, please look up the relevant machine parameters (including processor speed and network latency and bandwidth) to build a timing model for the efficiency. Assume you divide the function in a $1024 \times 1024 \times 1024$ mesh, please estimate the parallel efficiency by a BG/Q with $P = 2^{10}, 2^{12}, 2^{14}, 2^{16}$ processors.

Project C.24 Word Analysis

Please write a parallel program to search for the 10 most frequently used words in the first four chapters of Charles Dickens' *A Tale of Two Cities*. Please sort your list.

Project C.25 Cost Estimate of a 0.1 Pflops System

Supercomputers are designed to solve problems in science, engineering, and finance. Make a list of at least 10 major applications in those three areas and state the key requirements of the supercomputer resources in terms of computing power, network, memory, storage, and etc. If you are "commissioned" to "design" a supercomputer for the applications in one particular area, specify the key features of the system in the design. Try to get the prices of the components on the web and estimate the cost of the computer that can deliver 0.1 Pflops.

Project C.26 Design of a Pflops System

You are "commissioned" to design a 1-Petaflop "IntelGene" supercomputer with Intel's newest multi-core processors and the latest (2011) BlueGene network processors. In the design, you will need to specify the node architecture, the towers, and the final system. The terminology and packaging pathway of the published BlueGene design reports can be mimicked. The datasheet for the supercomputer's specifications in mechanical, thermo, electrical, and performance are to be provided.

APPENDIX D

PROGRAM EXAMPLES

D.1. Matrix-Vector Multiplication

Program D.1. Matrix-vector multiplication in FORTRAN.

```fortran
C multiplication of matrix M(N1,N2) and vector v(N1)
C product vector prod(N2)
C root process initializes M and v
C Broadcasts v to all processes
C splits M into sub_M by rows by MPI_Scatter
C (N2 is divided by the number of processes)
C each process multiplies its part of M x whole v and gets its part of
prod
C MPI_Gather collects all pieces of prod into one vector on root process
C simple case of N2 divisible by the number of process is handled here

      parameter(N1=8)        ! columns
      parameter(N2=8)        ! rows
      include '/net/campbell/04/theory/lam_axp/h/mpif.h'
      real M(N1,N2),v(N1),prod(N2)
      integer size,my_rank,tag,root
      integer send_count, recv_count
      integer N_rows

      tag = 0
      root = 0
C here(not always) for MPI_Gather to work root should be 0

      call MPI_Init(ierr)
      call MPI_Comm_rank(MPI_comm_world,my_rank,ierr)
      call MPI_Comm_size(MPI_comm_world,size,ierr)
      if(mod(N2,size).ne.0)then
         print*, ''rows are not divisible by processes''
         stop
      end if

      if(my_rank.eq.root)then
         call initialize_M(M,N1,N2)
         call initialize_v(v,N1)
      end if
```

```
C Broadcasts v to all processes
      call MPI_Bcast(v,
     @                N1,
     @                MPI_REAL,
     @                root,
     @                MPI_Comm_world,
     @                ierr)


C splits M into M by rows by MPI_Scatter
C (N2 is divided by the number of processes)
      N_rows = N2 / size
      send_count = N_rows*N1
      recv_count = N_rows*N1
C     if(my_rank.eq.root)send_count = N1*N2
      call MPI_Scatter(M,
     @                send_count,
     @                  MPI_REAL,
     @                M,
     @                  recv_count,
     @                  MPI_REAL,
     @                  root,
     @                MPI_COMM_WORLD,
     @                ierr)

      call multiply(prod,v,M,N_rows,N1)

      send_count = N_rows
      recv_count = N_rows
C     if(my_rank.eq.root)recv_count = N2
      call MPI_Gather(prod,
     @                  send_count,
     @                  MPI_REAL,
     @                prod,
     @                  recv_count,
     @                  MPI_REAL,
     @                  root,
     @                MPI_COMM_WORLD,
     @                ierr)

      if(my_rank.eq.root)call write_prod(prod,N2)
      call MPI_Finalize(ierr)
      end

      subroutine multiply(prod,v,M,N2,N1)
      real M(N2,N1),prod(N2),v(N1)

      do i=1,N2
       prod(i)=0
       do j=1,N1
```

```
      prod(i)=prod(i) + M(j,i)*v(j)
     end do
    end do
    return
    end

    subroutine initialize_M(M,N2,N1)
    real M(N2,N1)

    do i=1,N2
     do j=1,N1
       M(j,i) = 1.*i/j
     end do
    end do
    return
    end

    subroutine initialize_v(v,N1)
    real v(N1)

    do j=1,N1
      v(j) = 1.*j
    end do
    return
    end

    subroutine write_prod(prod,N2)
    real prod(N2)
C . directory for all process except the one the program was started on
C is your home directory

    open(unit=1,file='~/LAM/F/prod',status='new')

    do j=1,N2
       write(1,*)j,prod(j)
    end do
    return
    end
```

D.2. Long Range N-body Force

Program D.2.   Long-ranged N-body force calculation in FORTRAN (Source: Oak Ridge National Laboratory).

```
c  This program finds the force on each of a set of particles
interacting
c  via a long-range 1/r**2 force law.
```

```
c
c  The number of processes must be even, and the total number of points
c  must be exactly divisible by the number of processes.
c
c  This is the MPI version.
c
c  Author: David W. Walker
c  Date:    March 10, 1995
c
      program nbody
   implicit none
   include 'mpif.h'
   integer myrank, ierr, nprocs, npts, nlocal
   integer pseudohost, NN, MM, PX, PY, PZ, FX, FY, FZ
   real G
   parameter (pseudohost = 0)
      parameter (NN=10000, G = 1.0)
   parameter (MM=0, PX=1, PY=2, PZ=3, FX=4, FY=5, FZ=6)
      real dx(0:NN-1), dy(0:NN-1), dz(0:NN-1)
   real dist(0:NN-1), sq(0:NN-1)
      real fac(0:NN-1), tx(0:NN-1), ty(0:NN-1), tz(0:NN-1)
      real p(0:6,0:NN-1), q(0:6,0:NN-1)
   integer i, j, k, dest, src
   double precision timebegin, timeend
   integer status(MPI_STATUS_SIZE)
   integer newtype
   double precision ran
   integer iran
c
c  Initialize MPI, find rank of each process, and the number of
processes
c
      call mpi_init (ierr)
      call mpi_comm_rank (MPI_COMM_WORLD, myrank, ierr)
   call mpi_comm_size (MPI_COMM_WORLD, nprocs, ierr)
c
c  One process acts as the host and reads in the number of particles
c
   if (myrank .eq. pseudohost) then
         open (4,file='nbody.input')
      if (mod(nprocs,2) .eq. 0) then
         read (4,*) npts
         if (npts .gt. nprocs*NN) then
            print *,'Warning!! Size out of bounds!!'
         npts = -1
            else if (mod(npts,nprocs) .ne. 0) then
         print *,'Number of processes must divide npts'
         npts = -1
            end if
         else
```

```
      print *, ''Number of processes must be even''
      npts = -1
   end if
end if
```
c
c The number of particles is broadcast to all processes
c
```
call mpi_bcast (npts, 1, MPI_INTEGER, pseudohost,
#                   MPI_COMM_WORLD, ierr)
```
c
c Abort if number of processes and/or particles is incorrect
c
```
if (npts .eq. -1) go to 999
```
c
c Work out number of particles in each process
c
```
nlocal   = npts/nprocs
```
c
c The pseudocode hosts initializes the particle data and sends each
c process its particles.
c
```
    if (myrank .eq. pseudohost) then
    iran = myrank + 111
    do i=0,nlocal-1
       p(MM,i) = sngl(ran(iran))
       p(PX,i) = sngl(ran(iran))
       p(PY,i) = sngl(ran(iran))
       p(PZ,i) = sngl(ran(iran))
       p(FX,i) = 0.0
       p(FY,i) = 0.0
       p(FZ,i) = 0.0
       end do
    do k=0,nprocs-1
       if (k .ne. pseudohost) then
          do i=0,nlocal-1
          q(MM,i) = sngl(ran(iran))
          q(PX,i) = sngl(ran(iran))
          q(PY,i) = sngl(ran(iran))
          q(PZ,i) = sngl(ran(iran))
          q(FX,i) = 0.0
          q(FY,i) = 0.0
          q(FZ,i) = 0.0
             end do
          call mpi_send (q, 7*nlocal, MPI_REAL,
#                        k, 100, MPI_COMM_WORLD, ierr)
        end if
        end do
      else
      call mpi_recv (p, 7*nlocal, MPI_REAL,
#                  pseudohost, 100, MPI_COMM_WORLD, status, ierr)
```

```
      end if
c
c  Initialization is now complete. Start the clock and begin work.
c  First each process makes a copy of its particles.
c
      timebegin = mpi_wtime ()
      do i= 0,nlocal-1
          q(MM,i) = p(MM,i)
          q(PX,i) = p(PX,i)
          q(PY,i) = p(PY,i)
          q(PZ,i) = p(PZ,i)
          q(FX,i) = 0.0
          q(FY,i) = 0.0
          q(FZ,i) = 0.0
      end do
c
c  Now the interactions between the particles in a single process are
c  computed.
c
      do i=0,nlocal-1
        do j=i+1,nlocal-1
          dx(i) = p(PX,i) - q(PX,j)
          dy(i) = p(PY,i) - q(PY,j)
          dz(i) = p(PZ,i) - q(PZ,j)
          sq(i) = dx(i)**2+dy(i)**2+dz(i)**2
          dist(i) = sqrt(sq(i))
          fac(i) = p(MM,i) * q(MM,j) / (dist(i) * sq(i))
          tx(i) = fac(i) * dx(i)
          ty(i) = fac(i) * dy(i)
          tz(i) = fac(i) * dz(i)
          p(FX,i) = p(FX,i)-tx(i)
          q(FX,j) = q(FX,j)+tx(i)
          p(FY,i) = p(FY,i)-ty(i)
          q(FY,j) = q(FY,j)+ty(i)
          p(FZ,i) = p(FZ,i)-tz(i)
          q(FZ,j) = q(FZ,j)+tz(i)
        end do
      end do
c
c  The processes are arranged in a ring. Data will be passed in an
C  anti-clockwise direction around the ring.
c
    dest = mod (nprocs+myrank-1, nprocs)
    src  = mod (myrank+1, nprocs)
c
c  Each process interacts with the particles from its nprocs/2-1
c  anti-clockwise neighbors. At the end of this loop p(i) in each
c  process has accumulated the force from interactions with particles
c  i+1, ...,nlocal-1 in its own process, plus all the particles from
its
```

```
c   nprocs/2-1 anti-clockwise neighbors. The "home" of the q array is
C   regarded as the process from which it originated. At the end of
c   this loop q(i) has accumulated the force from interactions with
C   particles 0,...,i-1 in its home process, plus all the particles from
C   the nprocs/2-1 processes it has rotated to.
c
        do k=0,nprocs/2-2
        call mpi_sendrecv_replace (q, 7*nlocal, MPI_REAL, dest, 200,
     #                             src, 200, MPI_COMM_WORLD, status, ierr)
          do i=0,nlocal-1
            do j=0,nlocal-1
              dx(i) = p(PX,i) - q(PX,j)
              dy(i) = p(PY,i) - q(PY,j)
              dz(i) = p(PZ,i) - q(PZ,j)
              sq(i) = dx(i)**2+dy(i)**2+dz(i)**2
              dist(i) = sqrt(sq(i))
              fac(i) = p(MM,i) * q(MM,j) / (dist(i) * sq(i))
              tx(i) = fac(i) * dx(i)
              ty(i) = fac(i) * dy(i)
              tz(i) = fac(i) * dz(i)
              p(FX,i) = p(FX,i)-tx(i)
              q(FX,j) = q(FX,j)+tx(i)
              p(FY,i) = p(FY,i)-ty(i)
              q(FY,j) = q(FY,j)+ty(i)
              p(FZ,i) = p(FZ,i)-tz(i)
              q(FZ,j) = q(FZ,j)+tz(i)
            end do
          end do
        end do
c
c   Now q is rotated once more so it is diametrically opposite its home
c   process. p(i) accumulates forces from the interaction with particles
c   0,..,i-1 from its opposing process. q(i) accumulates force from the
c   interaction of its home particles with particles i+1,...,nlocal-1 in
c   its current location.
c
        if (nprocs .gt. 1) then
        call mpi_sendrecv_replace (q, 7*nlocal, MPI_REAL, dest, 300,
     #                             src, 300, MPI_COMM_WORLD, status, ierr)
          do i=nlocal-1,0,-1
            do j=i-1,0,-1
              dx(i) = p(PX,i) - q(PX,j)
              dy(i) = p(PY,i) - q(PY,j)
              dz(i) = p(PZ,i) - q(PZ,j)
              sq(i) = dx(i)**2+dy(i)**2+dz(i)**2
              dist(i) = sqrt(sq(i))
              fac(i) = p(MM,i) * q(MM,j) / (dist(i) * sq(i))
              tx(i) = fac(i) * dx(i)
              ty(i) = fac(i) * dy(i)
              tz(i) = fac(i) * dz(i)
```

```
            p(FX,i) = p(FX,i)-tx(i)
            q(FX,j) = q(FX,j)+tx(i)
            p(FY,i) = p(FY,i)-ty(i)
            q(FY,j) = q(FY,j)+ty(i)
            p(FZ,i) = p(FZ,i)-tz(i)
            q(FZ,j) = q(FZ,j)+tz(i)
          end do
        end do
c
c  In half the processes we include the interaction of each particle
with
c  the corresponding particle in the opposing process.
c
        if (myrank .lt. nprocs/2) then
          do i=0,nlocal-1
            dx(i) = p(PX,i) - q(PX,i)
            dy(i) = p(PY,i) - q(PY,i)
            dz(i) = p(PZ,i) - q(PZ,i)
            sq(i) = dx(i)**2+dy(i)**2+dz(i)**2
            dist(i) = sqrt(sq(i))
            fac(i) = p(MM,i) * q(MM,i) / (dist(i) * sq(i))
            tx(i) = fac(i) * dx(i)
            ty(i) = fac(i) * dy(i)
            tz(i) = fac(i) * dz(i)
            p(FX,i) = p(FX,i)-tx(i)
            q(FX,i) = q(FX,i)+tx(i)
            p(FY,i) = p(FY,i)-ty(i)
            q(FY,i) = q(FY,i)+ty(i)
            p(FZ,i) = p(FZ,i)-tz(i)
            q(FZ,i) = q(FZ,i)+tz(i)
          end do
        endif
c
c  Now the q array is returned to its home process.
c
        dest = mod (nprocs+myrank-nprocs/2, nprocs)
        src  = mod (myrank+nprocs/2, nprocs)
        call mpi_sendrecv_replace (q, 7*nlocal, MPI_REAL, dest, 400,
     #                        src, 400, MPI_COMM_WORLD, status, ierr)
        end if
c
c  The p and q arrays are summed to give the total force on each particle.
c
        do i=0,nlocal-1
          p(FX,i) = p(FX,i) + q(FX,i)
          p(FY,i) = p(FY,i) + q(FY,i)
          p(FZ,i) = p(FZ,i) + q(FZ,i)
        end do
c
c  Stop clock and write out timings
```

```
c
      timeend = mpi_wtime ()
      print *,'Node', myrank,' Elapsed time: ',
     #           timeend-timebegin,' seconds'
c
c  Do a barrier to make sure the timings are written out first
c
      call mpi_barrier (MPI_COMM_WORLD, ierr)
c
c  Each process returns its forces to the pseudohost which prints them
out.
c
      if (myrank .eq. pseudohost) then
         open (7,file='nbody.output')
      write (7,100) (p(FX,i),p(FY,i),p(FZ,i),i=0,nlocal-1)
      call mpi_type_vector (nlocal, 3, 7, MPI_REAL, newtype, ierr)
      call mpi_type_commit (newtype, ierr)
      do k=0,nprocs-1
         if (k .ne. pseudohost) then
            call mpi_recv (q(FX,0), 1, newtype,
     #                 k, 100, MPI_COMM_WORLD, status, ierr)
            write (7,100) (q(FX,i),q(FY,i),q(FZ,i),i=0,nlocal-1)
         end if
        end do
      else
       call mpi_type_vector (nlocal, 3, 7, MPI_REAL, newtype, ierr)
       call mpi_type_commit (newtype, ierr)
       call mpi_send (p(FX,0), 1, newtype,
     #             pseudohost, 100, MPI_COMM_WORLD, ierr)
        end if
c
c  Close MPI
c
  999   call mpi_finalize (ierr)

        stop
  100   format(3e15.6)
        end
```

D.3. Integration

Program D.3. Simple integration in Fortan (Souce: Oak Ridge National Lab).

```
program integrate
include "mpif.h"
parameter (pi=3.141592654)
integer rank
```

```
call mpi_init (ierr)
call mpi_comm_rank (mpi_comm_world, rank, ierr)
call mpi_comm_size (mpi_comm_world, nprocs, ierr)
if (rank .eq. 0) then
   open (7, file=''input.dat'')
   read (7,*) npts
end if
call mpi_bcast (npts, 1, mpi_integer, 0, mpi_comm_world, ierr)
nlocal = (npts-1)/nprocs + 1
nbeg   = rank*nlocal + 1
nend   = min (nbeg+nlocal-1,npts)
deltax = pi/npts
psum   = 0.0
do i=nbeg,nend
   x = (i-0.5)*deltax
   psum = psum + sin(x)
end do
call mpi_reduce (psum, sum, 1, mpi_real, mpi_sum, 0,
#                mpi_comm_world, ierr)
 if (rank.eq. 0) then
    print *,'The integral is', sum*deltax
 end if
 call mpi_finalize (ierr)
 stop
 end
```

REFERENCES

[1] S. H. Bokhari, On the mapping problem, *IEEE Transactions on Computers* **30**(3) (1981) 207–214.

[2] Top500 Supercomputing Sites. [Online]. HYPERLINK "file:///C:\\Users\\ zhlou\\Documents\\www.top500.org" www.top500.org.

[3] Y. Deng, A. Korobka, Z. Lou and P. Zhang, Perspectives on petascale processing, *KISTI Supercomputer* **31** (2008) 36.

[4] A. Grama, G. Karypis, V. Kumar and A. Gupta, *Introduction to Parallel Computing*, 2nd edn. (Addison–Wesley, 2003).

[5] W. Gropp, E. Lusk and A. Skjellum, *Using MPI: Portable Parallel Programming with the Message Passing Interface*, 2nd edn.(MIT Press, 1999).

[6] Y. Censor and S. A. Zenios, *Parallel Optimization: Theory, Algorithms, and Applications* (Oxford University Press, 1998).

[7] A. Gara *et al.*, Overview of the Blue Gene/L System Architecture, *IBM J. Res. Dev.* **49**(2–3) (2005) 195.

[8] A. Pacheco, *Parallel Programming with MPI*, 1st edn. (Morgan Kaufmann, 1996).

[9] G. Karniadakis and R. M. Kirby, *Parallel Scientific Computing in C++ and MPI* (Cambridge University Press, 2003).

[10] G. S. Almasi and A. Gottlieb, *Highly Parallel Computing* (Benjamin-Cummings Publishers, 1989).

[11] J. L. Hennessy and D. A. Patterson, *Computer Architecture: A Quantitative Approach*, 3rd edn. (Morgan Kaufmann, 2002).

[12] D. E. Culler, J. P. Singh and A. Gupta, *Parallel Computer Architecture — A Hardware/Software Approach* (Morgan Kaufmann, 1999).

[13] G. Amdahl, The validity of the single processor approach to achieving large-scale computing capabilities, *Proceedings of AFIPS Spring Joint Computer Conference*, Atlantic City, N.J., 1967, pp. 483–485.

INDEX